D0064857

The Social Leap

The Social Leap

THE NEW EVOLUTIONARY SCIENCE OF WHO WE ARE, WHERE WE COME FROM, AND WHAT MAKES US HAPPY

William von Hippel

HARPER WAVE
An Imprint of HarperCollins*Publishers*

HarperCollins books may be purchased for educational, business, or sales promotional use. For information, please email the Special Markets Department at SPsales@harpercollins.com.

FIRST EDITION

Designed by Fritz Metsch

Library of Congress Cataloging-in-Publication Data has been applied for.

ISBN 978-0-06-274039-7

18 19 20 21 22 LSC 10 9 8 7 6 5 4 3 2 1

To my father,

my first and best science teacher,

and my mother,

whose 50+year career inspired by example.

Contents

The Social Leap

Prologue

One morning when my son was eight years old, we decided to go sandboarding on Moreton Island, a small island made up entirely of sand that sits across the bay from our home in Brisbane. We arrived by ferry in the early afternoon and walked down the beach from the landing until we found a path through the forest to the massive dunes in the center. I had rigged up an old snowboard so my son could ride it barefoot, and once he found his balance, he was having the time of his life (in no small part because I was the one carrying the board up and he was the one riding it down). Climbing giant sand dunes is hard work, but the sun had well and truly set before I could convince him to call it quits.

He was chatty and happy as we walked back across the open dunes in the starlight, but the moment we reentered the woods, his mood changed. We could barely see the path ahead of us, and the forest that had felt so innocuous earlier now closed in around us. I could hear his voice start to quaver, and he quickly lost his train of thought. When a branch beneath my feet made a loud pop, he nearly jumped out of his skin. I tried to reassure him, but he insisted we were being hunted by wild animals. There was nothing I could say to dispel his fear; he was convinced a pack of dingoes was going to jump out at any moment to eat us. I have to admit, I felt a sense of dread as well, even though I knew our only real risk was a twisted ankle on the dimly lit forest trail.

Why did his happiness turn to fear so quickly? And why did I feel it too, even though I knew full well that the mosquitoes were the only animals that would be feasting on us that night? Perhaps surprisingly, the answers to these questions lie in the perceptual abilities of our distant ancestors. Humans have superb eyes but rather ordinary ears and noses, so other animals can detect us far more readily in the darkness than we can detect them. Our ancestors were fierce predators during the day, but at night they were prey, and nocturnal beasts have spent the last few million years making a meal out of any of our would-be ancestors foolish enough to go out at night. Those potential ancestors who wandered the woods in the moonlight were less likely to survive and procreate, and thereby less likely to pass on their proclivity for midnight strolls. This is how evolution shapes our psychology, with the end result being that no one needs to tell you to be afraid of the dark; it comes naturally.

If you go to the ape exhibit at your local zoo and spend some time with the chimpanzees, you can almost see evolution in action. They look like the distant cousins they are, and the ways they differ from us make perfect sense. It's not hard to see how leaving the forest could have caused legs like theirs to evolve into ours. Nor is it hard to imagine how evolution could have slowly transformed a second pair of hands into feet once our ancestors stopped climbing trees and started making long journeys on two legs.

What's less obvious is the role that evolution played in shaping our psychology. We tend to think of evolution in terms of anatomy, but attitudes are just as important for survival as body parts. Preferences that don't fit your abilities are as debilitating as limbs that don't suit your lifestyle. Our bodies changed a little over the last six or seven million years, but our psychology changed a lot.

Indeed, our evolution away from chimpanzees is marked primarily by adaptations to our mind and brain.

The most important changes in our psychology concern our social functioning, particularly our capacity to work together. By way of example, consider what happens when chimpanzees hunt monkeys. Monkey hunts are one of their few group-level activities because monkeys have much more difficulty escaping when chimps come at them from all sides. But even when chimps hunt as a group, not every chimp gets involved. Some sit idly by and watch the chaos around them. When the hunt is over, a few lucky chimps may have grabbed their prey, but most will be empty-handed. Meat is calorie-dense food, so the chimps who missed out on a monkey typically harass the chimps who caught one into sharing some of it. No surprise there. But what's notable is that chimps who only watch the hunt are just as likely to end up with a monkey snack as chimps who join the hunting party. Their fellow chimps make little or no distinction between slackers and helpers.

In sharp contrast, even children as young as four are attentive to who helps and who doesn't. When children earn stickers or candies by working as a team, they withhold their goodies from children who didn't help but share with children who did. This might not seem very friendly—it might even seem like behavior you should correct: sharing is caring, after all—but from an evolutionary standpoint, it's mission critical. Animals who don't distinguish between cooperators and bystanders will never have the capacity to create and maintain effective teams.

We tend to think of animals who live in groups as team players, but many animals live in large groups despite having very little engagement with one another. Wildebeest and zebra gather in huge numbers for safety, but they don't really show signs of teamwork. In a large group, someone else is likely to notice the lions, so each individual can afford to be a little less alert. Chimpanzees are much more interdependent than wildebeest or zebra, but even their lives rarely require genuine teamwork. As a consequence, they have limited capacity for cooperation and prefer to work alone. In contrast, once we left the trees, our very existence depended on our ability to work together. As we will see, our psychology was shaped by this need more than any other.

When our ancestors were expelled from the safety of the rainforest, they struggled to survive in the unknown and dangerous world of the savannah. Smaller, slower, and weaker than many of the grassland predators, they would have been doomed had they not happened upon a social solution to their problems. This solution was so effective that it put us on an entirely new evolutionary pathway. Our ancestors grew ever more clever precisely because they could leverage their newfound cooperative abilities to develop better ways to protect themselves and make a living. Eventually *Homo sapiens* became so smart that we started changing our environment to fit our own plans, most notably with the invention of agriculture. Farming hardened our hearts (and ruined our teeth), but it also allowed literature, commerce, and science to blossom.

Just because we got smarter doesn't mean we got any wiser. For better or worse, we haven't been able to shake many of our ancient instincts. Most notably, our fear of getting left out of the mating game still guides our psychology in profound ways, making us keenly aware of how we stack up compared to others in our group. This incessant social comparison is more disruptive to human happiness than almost anything else. It makes us nosy, too.

The ghosts of our evolutionary past continue to haunt us, but they also help answer some of the most fundamental questions about human nature. For example, how does the sociality we evolved on the savannah explain our ability and proclivity to innovate? What impact does it have on the way we lead and whom we follow? And how does it explain our regrettable tendency toward tribalism and prejudice? Our adaptation to life on the savannah may be ancient history, but it gives us new purchase on these modern problems.

Although we suffer from many of our ancestors' bad habits, they also evolved a motivational system that continues to reward us when we get it right. This is happiness. As is apparent in our fear of the dark, our motivations evolved to help us survive and thrive. That means that bad feelings serve an important purpose, but so do good ones. Our evolved psychology is deeply entwined with happiness and its pursuit; living the good life is largely a matter of meeting our evolutionary imperatives. Because these imperatives are often at cross purposes with one another, happiness is also a matter of figuring out how to navigate among them. Understanding the pressures exerted by our past can help guide us on this journey and can clarify why there are so many pitfalls along the way.

How Do We Know What Our Distant Ancestors Thought and Did?

Our deep past is called prehistory for a reason; there are no written records from the time period. Scientists have found an extraordinary number of fossils and other bits of evidence from our distant past, but sometimes these pieces of the past are open to multiple interpretations. Additionally, because strategies and behaviors don't fossilize, it's difficult to know exactly how our ancestors solved many of the problems they faced on their way to becoming human. Despite these challenges, evolutionary scientists have done a remarkable job extracting information from small clues, and their brilliant ideas and hard work have enabled me to tell this relatively complete story.

So how *do* we know what we know? To answer this question, let's consider three different approaches to the study of our evolutionary history: (1) how lice DNA indicate when we invented clothing; (2) how church records reveal the importance of grandmothers; and (3) how ancient teeth suggest what our ancestors did to avoid inbreeding.

HOW DO WE KNOW WHEN WE INVENTED CLOTHING?

Humans have the distinct pleasure of being the host to three different species of lice: head lice, pubic lice, and body lice. The story of how we came to provide these revolting little parasites with a home that is also a meal is an intricate one, and it begins with the head lice my children brought home from day care. The ancestors of human head lice infested primates about twenty-five million years ago, which is around the time apes and Old World monkeys (i.e., monkeys from Africa and Asia) went their separate ways.

When our more immediate ancestors split from the ancestors of chimpanzees six or seven million years ago, the lice that accompanied us could roam anywhere on our bodies, as our ancestors were still a hairy lot. These ancient body lice were the only species that plagued us at the time, but a few million years later we caught a new species of lice, apparently from gorillas. I'm not sure how our ancestors managed that one, but I'd like to think they were just living in close proximity to gorillas, maybe sharing the same bed on occasion to stay warm. Whatever the cause, about three million years ago we began hosting two distinct species of lice.

As we continued down our evolutionary pathway, we eventually lost our thick body hair (and our habit of consorting with gorillas). Our newfound hairlessness posed a problem for both our species of lice, as they depend on a forest of hair to deposit their eggs. The end result was that we forced these two species of lice to become specialists. The lice that had accompanied us for the longest retreated to the northernmost part of our body and became head specialists. The lice we caught from gorillas moved to our equatorial region and became crotch specialists.

This détente between our two lice species remained in place for about a million years, until just seventy thousand years ago, when a third species of louse appeared on the scene, an offshoot of the lice on our heads. These new lice evolved to live on our body, but just like the lice from which they originated, they couldn't lay their eggs on our (now-hairless) skin, as the eggs would have fallen to the ground and died. Rather, these new lice required clothing to deposit their eggs. For this reason, the evolution of body lice provides pretty good evidence that we started wearing clothing by at least seventy thousand years ago.

Of course, the tricky questions are why did we bother with clothes, and why then? Our ancestors had been hairless for over a million years at that point, and most of them still lived in the warm climate of Africa—but not all of them. As we will see, just prior to the advent of body lice, *Homo sapiens* had begun migrating out of Africa. Perhaps this migration to colder climates led to the invention of clothing. Or perhaps clothing was invented much earlier and was intended to shield us from sun as well as cold. Alternatively, perhaps our ancestors were just seeking to ornament themselves or differentiate themselves from others. Whatever the reason, from that point forward at least some of our ancestors must have worn clothing most of the time, or our body lice would have died out.

The evolutionary story of body lice provides great evidence regarding the invention of clothing, but how do we know the details of this time line? And how do we know we got our pubic lice from ancestral gorillas three million years ago? To answer such questions, scientists have relied on molecular clocks, which are timing procedures based on DNA mutation rates. Once two species diverge, they start to build up random mutations in their DNA. These mutations are no longer shared between the two species, and hence are unique to each. Because we know the average pace of mutation on different strands of DNA, we can count the unique mutations on strands of DNA that are shared by both species to assess when the two species went their separate ways.

For example, if a particular strand of DNA in a particular species mutates at an average rate of once every twenty generations, and if we find an average of fifty distinct mutations on this DNA in each of two previously related species, we know that they have been separated for about a thousand generations. When we count backward in this way, we eventually get to the parent species that is genetically closest to the two offspring species.

By studying mutation counts in the DNA of body lice and head lice (which are closely related to each other but not to pubic lice), we have pretty good evidence that our ancestors stopped running around naked at least seventy thousand years ago. Using this same procedure, we also have pretty good evidence that our pubic lice have been separated from gorilla lice for about three million years.

HOW DO WE KNOW IF GRANDMOTHERS ARE IMPORTANT?

The Lutheran Church has maintained meticulous records of all births, marriages, and deaths in Finland since the eighteenth century. Mirkka Lahdenperä of the University of Turku and her colleagues took advantage of this excellent source of data to plot the life course of more than five hundred women and their children and grandchildren from five different farming and fishing communities in Finland between 1702 and 1823.

By carefully combing through these records, Lahdenperä and her colleagues discovered several important facts about grandparents. Perhaps most remarkably, they found that for every ten years a grandmother lived beyond the age of fifty, she gained two extra living grandchildren. This effect emerged most clearly in families in which grandparents lived in the same village as their grandchildren, and seemed to be a function of three factors:

1. A living grandmother in the same village enabled daughters to start having their own children earlier (at an average age of 25.5 versus 28).

2. A living grandmother also shortened the interval between births, as daughters of living grandmothers had children every 29.5 months, but daughters of deceased grandmothers had children every 32 months.

3. A living grandmother who was under the age of 60 (and thus more likely to be energetic and helpful) increased survival rates of grandchildren by 12 percent. This increased survival rate manifested itself only post-weaning, as children who were still being breast-fed survived at similar rates whether their grandmother was alive or not.

During this period in Finland (and everywhere else), illnesses and injuries took nearly half of the children before they reached adulthood, so these positive effects of grandmothers on survival and reproduction were keenly felt.

HOW DO WE KNOW WHAT OUR ANCESTORS DID TO AVOID INBREEDING?

Animals that live in small groups gain numerous advantages from group living, but they face a problem of how to avoid inbreeding. Without knowledge of their family tree, animals that are born into small groups and then mate with members of that group risk mating with close relatives.

There are several potential costs to mating with close relatives, but the most notable one is that dangerous genes are more likely to find a match when you mate inside the family. For example, I carry a gene for Tay-Sachs disease, which fortunately for me is recessive (meaning that unless you get the Tay-Sachs gene from both parents, you suffer no consequences). When both parents carry the Tay-Sachs gene, there is a 25 percent chance that each of their children will get two Tay-Sachs genes and suffer from the disease. Most Tay-Sachs victims show signs of the disease by six months of age, at which point they begin to lose their sight and hearing, then their ability to swallow, and eventually their ability to move, and soon afterward they die.

The gene for Tay-Sachs is rare (fewer than one out of every two hundred people carry it in the general population), so there is almost no risk that carriers like me will have a child who has Tay-Sachs because there is almost no chance that they'll happen to fall in love with a fellow Tay-Sachs carrier. But if I were to have children with members of my family, such as my siblings or cousins, there would be a much greater likelihood that my partner carried the same Tay-Sachs gene I do, and a much greater likelihood that our children would suffer from this horrible disease.

The most common way that animals who live in small groups solve this potential inbreeding problem is by having *either* males or females leave the group in which they were born when they reach adolescence. By leaving their group behind and joining a new one, animals dramatically reduce the likelihood of mating with someone who is a close relative. It's important to keep in mind, however, that animals have no idea why they leave their group. Rather, those animals who developed a wanderlust and migrated to a new group were more likely to avoid these inbreeding problems. As a consequence, the tendency to change groups spread through the species via the enhanced reproductive success of animals who inherited the tendency to wander off when they reached sexual maturity.

Chimpanzees solve this inbreeding problem by having the females find new groups when they reach maturity. In contrast, hunter-gatherer humans are more flexible and varied in their solutions (more on this issue in chapter 3). Researchers wondered whether our distant ancestors were similar to chimps in this regard or more similar to us. But how do you piece together that sort of information when all you have are random bits of fossils, with nothing else that survived to tell the tale of how our ancestors lived?

Scientists cracked this particular nut by measuring strontium levels in our ancestors' teeth. Strontium is a metal that is absorbed into the body in a manner similar to calcium, and hence it can be found primarily in our bones and teeth. There are four different forms of strontium, and the ratio of these different forms varies with the local geology. Some locations have strontium that is very common in one form, relatively common in another, and rare in the remaining two; and other locations have different patterns.

Because strontium is incorporated into the teeth during growth and development, ancient teeth can be analyzed to assess the ratio of different forms of strontium. If the strontium ratio found in ancient teeth matches the ratio found in the local bedrock, whoever owned the teeth almost assuredly grew up in the region where their teeth were found. In contrast, if the ratio differs from the local bedrock, the owner of the teeth almost assuredly moved to that region after childhood.

When Sandi Copeland of the Max Planck Institute for Evolutionary Anthropology and her colleagues analyzed the strontium ratios from the teeth of various *Australopithecus africanus* (our ancestors from a few million years ago; more on them in chapters 1 and 2), she found that the larger teeth matched the local geology but the smaller teeth did not. Because males are typically larger than females, and hence have larger teeth, these data suggest that female *Australopithecines* probably left the groups in which they were born, and thus avoided inbreeding, just like chimpanzees.

As you can tell from these three lines of research, scientists use a variety of approaches to study our past. Sometimes the data give us a lot of confidence in our conclusions, such as when we see that grandmothers living in the same town are associated with reduced childhood mortality. Other times the data allow educated guesses, such as when we infer that smaller teeth are female and thus females likely left their birth groups when they reached maturity. Still other times the data only provide constraints on our theorizing, such as when the emergence of body lice gives us the latest date by which we must have invented clothing but doesn't provide clear evidence about what the earliest date might be—perhaps lice took their sweet time in adapting to the newfound opportunities of clothing.

It's important to remember in this regard that any individual study is just a small piece of the puzzle; it's the combination of thousands of studies that provides us with the overall picture. When the studies all point in the same direction, we can be pretty sure we understand what's going on. When they contradict one another or have multiple interpretations, we have more work to do. Unsurprisingly, as we go further back in time, the evidence becomes thinner and more ambiguous, and we are forced to rely increasingly on conjecture. Be that as it may, I have tried to tell our story without the endless caveats that make academic writing tedious and difficult to read. So please keep in mind that this book represents my best effort to explain who we are and how we got here, based on the incomplete, complicated, and sometimes contradictory data that exist. For readers interested in learning more, I've included a reference section at the end of the book that is separated by chapter.

Nature versus Nurture?

I have one last point I'd like to make before diving into the book, which concerns the role of nature and nurture in our psychological makeup. Some people are offended by evolutionary approaches to human behavior, criticizing evolutionary psychology for what they perceive as its implications. Such people often believe that if genes influence the contents of our minds, those aspects of our minds that are subject to genetic influence are impervious to environmental or social influences and are outside personal control. I want to clarify that nothing could be further from the truth. By way of example, let's consider a body part that is much simpler than our brain: our muscles.

Differences in our genes give us the capacity to grow muscles of different sizes. Some people inherit a proclivity to grow large muscles (the front line on any major football team comes to mind), and some people inherit a tendency toward more modest musculature (if you knew me, I might come to mind). Our genes provide the blueprint that enables our muscles to grow to varying degrees when they are repeatedly overtaxed—for example, by weight training, manual labor, or participation in sports.

Nonetheless, it is our lifestyle that determines whether we subject our muscles to more or less strain or provide them with more or less nutrition, thereby causing them to grow or shrink. As a result, muscles of different size are a product of our genes, our environment, and the interaction between our genes and our environment. At the same time, our musculature can also be a matter of personal choice. As this example highlights, evolutionary theory conceives of neither body nor mind as the product of some sort of competition between nature and nurture, nor as the product of an inflexible biological program, nor as something removed from human agency and choice.

These interactions between genes and environment emerge even when genetic effects are very strong. For example, myopia (nearsightedness) is highly heritable, and nearsighted parents are likely to have nearsighted children. Yet studies of hunter-gatherer eyesight show that there are almost no nearsighted hunter-gatherers. There are various aspects of modern life that might cause myopia—perhaps it's all the close work we do, perhaps it's reading, perhaps it's working in low light—but whatever the cause, the genes that lead to myopia are actually genes that make people sensitive to environmental factors that cause myopia. People who have myopia genes and live in modern environments usually develop nearsightedness; people who have myopia genes but live as hunter-gatherers almost never do. So even effects that are largely genetic can at the same time be largely environmental.

This principle also holds true when it comes to our mind. The contents of our mind are a product of our genes, our environment, and our personal choices. Our genes nudge us in certain directions—sometimes this nudge might more aptly be described as a shove—but we make the decisions that determine the trajectory of our lives.

There are countless examples of human choice overwhelming genetic tendencies, but perhaps a life of celibacy is the clearest case of all. One of the strongest desires that our genes give us is the desire to have sex, because the absence of sex ensures that our genes end with us. Despite that fact, a large number of humans throughout history have decided to forgo all sexual activity. Many have struggled but failed to enact this decision, but many have succeeded. No doubt some of the successful ones wrestled mightily with their decision, but that's the point. Just because our genes give us a massive shove in their preferred direction doesn't mean we have to go that way.

It's easy to imagine a world in which genes have control over our minds, and for many animals they do. But once we took the evolutionary pathway toward greater intelligence and a lifestyle that relies on learning rather than inborn knowledge, our genes had no choice but to relinquish much of their control.

By way of example, consider how meerkats train their young to hunt. Meerkats get most of their nutrition by eating insects, and the ones who live in the Kalahari Desert can't be too fussy about which insects they eat. One of their prey animals is the scorpion, which is obviously a tricky dinner choice given that it has the power to kill in return. Meerkats are not born knowing how to kill a scorpion, so their parents and older siblings teach them.

As part of their teaching technique, meerkats differentiate how they bring home a scorpion meal as a function of how old the pups are. When the pups are newly weaned, the adult meerkat kills the scorpion before giving it to the pups. When the pups are larger, the adult meerkat breaks off the scorpion's stinger before handing it over but leaves the scorpion alive so the pups can practice killing it themselves. Finally, when the pups are getting ready to venture out on their own, the adult meerkat hands over a live and intact scorpion, which the pups must attack and kill for their dinner.

This process gives the impression of being well thought out, but meerkats will rely on just one signal to determine how to handle the scorpion before giving it to the pups: sound. When researchers play the sounds of very young pups, meerkats kill the scorpion before handing it over. When researchers play the sounds of older pups, the meerkats hand over a live and deadly scorpion. Amazingly, the sounds made by pups at different stages of development induce these behaviors in their adult carers *independent of how old the pups actually are.* Despite the fact that the carers are in direct daily contact with young and nearly helpless pups, they will offer them an intact scorpion if they hear calls that are made by older and more capable pups.

Data such as these show that meerkat decisions can be determined by the combination of their genes and just one piece of environmental information. No doubt this system evolved because it was computationally efficient (i.e., it didn't require too much brainpower), and in the real world it worked very well—baby pups never make adolescent sounds.

Humans stand in sharp contrast to meerkats and other animals like them. Our genes also influence our decisions, but only in combination with a huge range of input, some of which comes from inside our skull and is a function of how we see ourselves and who we want to be. For this reason, human agency remains an important determinant of behavior, as people *decide* whether they're going to be easygoing or forceful, cooperative or competitive, and ambitious or lazy. Our genes are one factor in that decision-making process, but they are only a single factor. As we saw with myopia, genes interact with the environment to exert their influence, so to acknowledge the power of genes is not to refute the importance of upbringing, social class, culture, and such.

The bottom line is that evolutionary psychology is a story about how evolution shaped our genes, which in turn sculpt our minds, but it is not a genetically deterministic story at all. The environment also sculpts our minds, and our culture, values, and preferences play a critical role in who we become—and where we go next.

Part I

How We Became Who We Are

1

Expelled from Eden

You and I are descendants of chimplike creatures[*] who left the rainforest and moved to the savannah six or seven million years ago. On first glance it would seem like an odd decision for our ancestors to leave the trees, as there were virtually no predators that could hunt them successfully when they were in the forest canopy. Even superb tree climbers such as leopards don't attack chimps in trees, as chimps are simply too fast and too dangerous when they are in their element. On the ground, however, chimps are easy prey. They are ungainly on two legs, comparatively slow on all four, and their small size makes them an easy meal for large cats such as lions, leopards, or the saber-toothed tigers that once roamed East Africa.

So why leave the trees? What compelled our ancestors to trade the safety and sheer exuberance of life in the canopy for a slow and clumsy existence on the ground? There is vigorous scientific debate on this question, but one widely endorsed theory is an updated version of the "savannah hypothesis." This hypothesis was proposed

* Humans and chimps evolved from a common ancestor, but we don't know exactly what that ancestor looked like. The fossil record strongly suggests that our shared grandparents looked far more like today's chimps than like us. For this reason, I refer to our common ancestors as *chimplike* or *chimpish.*

by Ray Dart in 1925, when he published the discovery of *Australopithecus africanus*, or "the man-ape of South Africa." After noting that humans were unlikely to have evolved in tropical forests because life there was too easy, Dart wrote, "For the production of man a different apprenticeship was needed to sharpen the wits and quicken the higher manifestations of the intellect—a more open veldt country where competition was keener between swiftness and stealth, and where adroitness of thinking and movement played a preponderating role in the preservation of the species."

Dart was right that we evolved in the savannah, but in 1925 he had no idea what forces put us there. We now believe that tectonic activity along the East African Rift Valley is what split us from our chimpish ancestors. All the earth's surfaces, including the landmasses that make up the continents and the bottoms of the oceans, sit on tectonic plates. These plates float around on an underlying mantle, which emerges as a viscous liquid when it flows from a volcano but is under so much pressure below the earth's crust that it is more like pliable road tar. The heat emanating from the earth's core creates incredibly slow but strong currents in the mantle, and these currents carry the plates around with them. Sometimes these plates ram into each other in super slo-mo, as is the case with India smashing into Asia, a by-product of which is the Himalayas (which continue to rise a few centimeters each year). Sometimes these plates tear apart and move away from one another. In Africa, the east side of the continent is slowly unzipping from the rest, starting at the Red Sea, in the north, and ending at the coast of Mozambique, in the south.

The tectonic activity along this geographic zipper created the East African Rift Valley and slowly and sporadically raised vast portions of Ethiopia, Kenya, and Tanzania to an elevated plateau. These changes in topography led to localized changes in climate, with the rainforests on the east side of the Rift Valley drying out one by one, to be replaced by savannah. So it turns out that we didn't leave the trees after all—the trees left us.

Because our chimpish ancestors were so impressive in the trees and so unimpressive on the ground, the gradual replacement of the rainforest with savannah meant that they had to find a new way to make a living. The fruits, berries, and leaf buds they were accustomed to eating receded along with the trees, their opportunities to hunt for meat were greatly diminished by their slow speed on the ground, and, to top it off, enormous predators prowled the grasslands. So how did our ancestors respond to this double whammy of disappearing food and newly dangerous predators? No doubt many of our would-be ancestors perished, but some of them survived and eventually began to thrive, and their story is our own.

The Dik-Dik/Baboon Strategy

Our chimpish ancestors are not the only tree dwellers who ever tried out life on the ground, so scientists often look to the behavior of other species to see how chimps might have adapted to the grasslands. One analogue can be found in baboons. Although baboons are monkeys and not apes (and hence not as clever as chimps), they resemble chimps in many ways, and several baboon species reside on the African savannah. Savannah baboons live in large groups, which gives them the advantages of many eyes to watch for predators and many teeth with which to defend themselves. The "baboon solution" to savannah life isn't a terrible one, as evidenced by the fact that there are still plenty of baboons, but it *is* stressful and fraught with danger. Baboons often meet an abrupt end in the mouth of a hungry lion or leopard.

In their confrontations with predators, baboons depend heavily on their massive incisors, which are larger than those of a chimp even though baboons themselves are smaller. If our chimpish ancestors had "decided" that biting was the answer to their savannah dilemma, our faces would likely be more doglike than they are today, with a protuberant jaw and much larger teeth. Our tiny jaws and pathetic canines indicate that the baboon solution doesn't appear to have suited our ancestors, who took a different approach to life on the plains. Indeed, this decision was already evident by the time we had evolved into Ray Dart's *Australopithecus*, whose jaw and teeth were halfway between a chimp's and our own.

Because chimps are brainier than baboons, they take longer to reach adulthood, and their slower maturation rate means they demand more maternal care. As a consequence, chimps have an older age of initial reproduction and a lower rate of reproduction than baboons. This slower reproduction would have put our ancestors at a greater risk of extinction if they had been picked off at the same frequency as baboons. For this reason, our chimpish ancestors who survived this evolutionary pressure cooker were probably the ones who did their utmost to escape the notice of lions, saber-toothed tigers, and other predators rather than taking a more confrontational approach.

Indeed, hiding is the primary survival strategy for many herbivores. Consider the dik-dik, an antelope about the size of a house cat that also lives on the East African savannah. By virtue of their diminutive size, dik-diks have no defense against any predator larger than a poodle, so they spend their lives hiding. They are impressively quick and agile when chased, but not fast enough to survive being hunted on the open grasslands. As such, dik-diks blend into their surroundings, stay on the lookout for predators, and never stray far from heavy bush.

Our chimpish ancestors weren't as quick as dik-diks, but they could climb trees. It's likely they spent their day hiding, watching for predators, and scrambling up nearby trees for safety. When modern chimps are in the savannah, they adopt this sort of combined dik-dik/baboon approach, clustering together more than chimps do in the rainforest and cautiously avoiding open areas where there are no trees available for a ready escape. Perhaps even more interesting, savannah chimps exhibit two other unique behaviors: they fashion crude spears out of tree branches, which they use to poke into tree hollows to skewer and retrieve the monkeys hiding inside, and they are more likely than rainforest chimps to share with one another. Both these behaviors mimic changes shown by our ancestors after they left the forest (more on this later).

These data from savannah chimps and baboons suggest that greater watchfulness would have allowed our ancestors to eke out a living on the savannah, and probably played an important role in their survival for the first few million years after the disappearance of the forest. Unlike baboons and dik-diks, however, our ancestors were not content with this modicum of success. The savannah brought with it new opportunities for a clever ape whose hands were no longer required for locomotion. Change didn't come overnight, but across the ensuing three million years, numerous adaptations to our minds and bodies suggest that we found entirely new ways to protect ourselves on the grasslands.

Throwing Rocks at Lions

What would you do if you were attacked by an animal that was too strong, too ferocious, and too fast for you to flee or fight off with your bare hands? In my case, it doesn't take much imagination to answer this question. I grew up in a neighborhood that was inattentive to leash laws, and my friends and I were often chased by a German shepherd and Doberman pinscher that lived on our street. Even though I was a scrawny kid, and these dogs would still intimidate me today, by the age of seven or eight I had become pretty good at defending myself by throwing stones. Especially if my brothers or friends were with me, all we had to do was bend over to gather rocks, and the dogs running toward us would pull an immediate about-face. When I was alone, I took off for the nearest fence or tree, because I couldn't throw rocks fast enough to do the job, but the addition of even one other person meant we could stand our ground.

These experiences suggest how our ancestors might have responded to the threat of predation on the savannah: by throwing stones, particularly if they could band together and throw lots of them. We can't look back in time to see if that's what they did, but we can look at differences between our bodies and theirs to see if this strategy is plausible. So, what does the evidence show?

Sure enough, a number of changes in the fossil record support the stone-throwing hypothesis. Most of these changes can be found at least partially in our ancestor *Australopithecus afarensis* (aka Lucy, who roamed East Africa three and a half million years ago and was a predecessor of Ray Dart's *Australopithecus africanus*). Lucy wasn't much brighter than a chimp, judging by the size of her brain, but she appears to have devised new ways to deal with predators beyond hiding and hoping not to be noticed. Compared to a chimpanzee, she had a more mobile hand and wrist, more flexibility in her upper arm, a more horizontally oriented shoulder, and more space between her hip and the bottom of her rib cage. This changing constellation of traits was likely a product of the fact that she was bipedal (she walked upright), a habit her ancestors evolved on the savannah. These new traits were also incredibly useful for throwing.

When you watch people toss a ball back and forth at the beach, you may get the impression that throwing is mostly a function of arm and shoulder muscles. If you want to learn to throw with power and accuracy, however, you need to watch baseball players, quarterbacks, or hunter-gatherers. Among experienced throwers, arms and shoulders are just a small part of the equation. Power throwing begins by stepping forward with the opposite-side leg (e.g., a left foot step for a right-hander), progresses through rotation of the hips, followed by rotation of the torso and then shoulders, and finally the elbow and wrist follow through.

These sequential motions take advantage of the fact that the combined forward and rotational forces of the body stretch the ligaments, tendons, and muscles of the arm and shoulder, which accelerate the arm forward at the very end of the throw, like the snapping of a rubber band. Chimps are stronger than we are, but they can't generate this sort of elastic energy when they throw because their joints aren't flexible enough and their muscles don't line up in the right way. These changes to the hips, shoulders, arms, wrists, and hands are what made Lucy and her fellow *Australopithecines* much better stone throwers. These same changes also supported excellent clubbing,* which would have been useful whenever throwing failed to do the job.

Driving off a Doberman with rocks is one thing; driving off lions and saber-toothed tigers is another challenge altogether, especially when you weigh between sixty and a hundred pounds and stand three and a half to five feet tall, as *Australopithecines* did.† Nonetheless, throwing can be incredibly effective if you practice a lot. I first had my nose rubbed in this fact when I was in my late twenties and visited the Ohio State Fair with my girlfriend. One of the stalls had a pitching net with a radar gun, and I decided to impress her with my athletic prowess. I was pretty pleased with my fifty-mile-per-hour throws, and she seemed suitably awed—until a gangly twelve-year-old set up shop next to me. Without so much as breaking a sweat, this prepubescent eighty-five-pounder easily hurled ball after ball at sixty-plus miles per hour. Not wanting to lose this manly contest to a human twig, I threw my last ball as hard as I possibly could and was rewarded with a wildly inaccurate

* That is, the thwacking variety of clubbing; there is no fossil evidence for an ancient party scene on the savannah.

† Which is a little taller but a little lighter than a modern chimp.

fifty-five-mile-per-hour pitch and excruciating pain in my elbow and shoulder. My girlfriend consoled me by suggesting that throwing was more practice than power—I think this was the moment when I first knew I wanted to marry her—and of course she was right.

Keeping in mind that practice makes perfect, we see that the throwing hypothesis is more plausible, particularly if throwing is taken up by an entire group. Consistent with this possibility, the historical record also indicates that throwing can be remarkably effective. There are numerous descriptions of encounters between European explorers and indigenous populations in which conflict ensued and the indigenous population was armed only with stones. The European explorers typically relied on guns and armor, but they often lost these skirmishes, sometimes badly. Consider these three historical accounts that anthropologist Barbara Isaac dug up for her wonderful article "Throwing and Human Evolution."

> In hardly any time at all they had so badly beaten us that they had driven us back into shelter with heads bloodied, arms and legs broken by blows from stones: because they know of no other weaponry, and believe me that they throw and wield a stone considerably more skillfully than a Christian; it seems like the bolt of a crossbow when they throw it.
>
> —*Jean de Béthencourt, 1482*

> The enormous stones hurled by the savages maimed one or other of our people at every moment . . . a shower of stones, so much the more difficult to avoid, as being thrown with uncommon force and address, they produced almost the same effect as our bullets, and had the advantage of succeeding one another with greater rapidity.
>
> —*Jean-François de Galoup de La Pérouse, 1799*

> Many a time, before the character of the natives was known, has an armed soldier been killed by a totally unarmed Australian. The man has fired at the native, who, by dodging about has prevented the enemy from taking correct aim, and then has been simply cut to pieces by a shower of stones, picked up and hurled with a force and precision that must be seen to be believed . . . the Australian will hurl one after the other with such rapidity that they seem to be poured from some machine; and as he throws them he leaps from side to side so as to make the missiles converge from different directions upon the unfortunate object of his aim.
>
> —*John Wood, 1870*

These accounts highlight the potential deadliness of collective stone throwing, but they also highlight a crucial point: cooperation is the key to making this strategy a success with large animals such as lions and leopards.

The Psychology of Collective Action

Chimps are more likely to compete with one another than they are to cooperate, and thus it would have been difficult for our chimplike distant ancestors to act collectively to drive off large predators. A lone *Australopithecus afarensis* throwing stones (perhaps while other members of its group ran away) would have ended up in the belly of a slightly bruised predator, but many *Australopithecines* throwing stones could probably have driven off hyenas, saber-toothed tigers, and even lions. It was this need for collective action that brought about the most important psychological change that enabled us to thrive, rather than just survive, on the savannah: the capacity and desire to work together.

Modern chimpanzees cooperate loosely with one another when they hunt as a group and when they attack other chimps as a group, but their fundamental orientation toward group members who are not kin or close friends is competitive. Thus, it is likely that the first hundred, thousand, or even million times our chimpish ancestors were sneaking across the grasslands, they scattered for the nearest trees at the first sign of attack. But somewhere along the line, our ancestors banded together in their collective defense, at which point they *all* stood a better chance of survival.

Individuals in groups who learned to work cooperatively in this manner were at an enormous advantage, and would have easily outbred individuals in groups committed to a strategy of "every chimpish chap for himself." Just as important, evolution would have favored any subsequent psychological changes that supported the quality of the group's collective response. Our ancestors who liked to cooperate, and who could be counted on by others to be cooperative, reaped a great reward as a result.

Once *Australopithecines* learned to fend off predators by throwing stones, they would have soon discovered that they could also hunt via collective stone throwing. Collective stone throwing requires little advance planning or coordination, and thus was possible even with the limited cognitive abilities of our distant ancestors. Whenever a group of *Australopithecines* happened upon possible prey, they would likely have pelted them with rocks. Throwing would also have enabled *Australopithecines* to scavenge recent kills from other animals, as any lone creature that had made a kill would soon join its prey in the pot if it tried to defend its dinner in the face of hurtling stones.

Stone throwing not only massively enhanced the benefits of cooperation but also created new means to enforce it. The greatest challenge to cooperation is free riding, or the tendency to skip the hard work while sharing the benefits. Many of the *Australopithecines* in these early savannah groups would have been tempted to be free riders, running away at the first sign of predators while the rest of the group worked cooperatively to fend them off. No doubt our ancestors found such free riders frustrating, just as we do when members of our work groups don't pull their weight but still nod wearily when our boss thanks the team for the all-nighter. But now our ancestors had new weapons at their disposal to ensure cooperation.

Their first weapon was the threat of ostracism. To be forced out of a group of apes in the forest was a bummer, but to be forced out of a group of *Australopithecines* on the grasslands was a death sentence. For this reason, our ancestors rapidly evolved a strong emotional reaction to the threat of being ostracized.* The *Australopithecines* who didn't mind being tossed out of their group are not the *Australopithecines* who became our ancestors, so the threat of being ostracized soon brought free riders in line. Ostracism and rejection have remained important tools for enforcing cooperation through to the present, and as a result we still find social rejection incredibly painful and do whatever it takes to stay in our group's good graces.

For repeat offenders who were difficult to ostracize (either because they stuck to the group like limpets or because they were aggressive and didn't take kindly to ostracism), the threat of collective punishment likely worked wonders. The ability to kill at a distance is the single most important invention in the history of warfare, because weaker individuals can attack stronger individuals from a position of superior numbers and relative safety. Stoning was probably one of the earliest forms of punishment our ancestors meted out to peers who failed to do their part, and it has remained a common punishment through to recent times. For example, the Bible invokes stoning as retribution for a variety of sins, even though it

* By *ostracized*, I mean the original (Greek) use of the term, which refers to banishment from one's group. Psychologists often use the term to refer to ignoring someone or treating them as if they weren't present. This use of the term represents a psychological form of banishment, which is certainly unpleasant, but has very different survival consequences from getting the heave-ho in an ancestral environment.

was written when people could also hang, decapitate, crucify, or otherwise kill one another in terribly inventive ways. The safety* and effectiveness of stoning transgressors were not lost on the creators of these biblical laws.

Although throwing rocks is not exactly rocket science, this early collective action sparked the evolutionary processes that led to the extraordinary expansion of our mental capacities over the ensuing three million years. The decision to throw rocks at predators might not seem like a big deal, and it might not have helped all that much the first few hundred or thousand times, but when it finally worked, it changed everything.

Collective Action Brought About the Cognitive Revolution

Scientists once believed that we became so smart to take advantage of the opportunities to manipulate objects that are afforded by opposable thumbs. There is undoubtedly some truth to this possibility; after all, octopi are awfully smart, and their tentacles provide the same opportunities as opposable thumbs. A huge brain would also be of little use to a zebra, who couldn't possibly make or wield any tools with its hooves.

Ultimately, though, dealing with fellow group members is a much greater mental challenge than manipulating objects. For this reason, many scientists have adopted the social brain hypothesis, which is the idea that primates evolved large brains to manage the social challenges inherent in dealing with other members of their highly interdependent groups.† This hypothesis has particular pur-

* For those doing the stoning; it was obviously hard on the recipient.

† Social challenges play an important role in the cognitive capacities of numerous other species as well. For example, elephants track the whereabouts of their group members over great distances, and magpies grow into more clever adults if they live in larger groups.

chase with humans, and not just because we live in larger groups than other great apes. Rather, once our ancestors began reaping the benefits of teamwork, they laid the groundwork for all sorts of social innovations, most of which came one or two million years later (and are the topic of chapter 2). These social innovations required a larger brain to coordinate and achieve, and they put greater pressure on our ancestors to get smarter.*

Cooperation made our ancestors smarter, but cooperation also demanded numerous changes in the ways their minds worked. First and foremost, our ancestors began to benefit from information sharing. In their previously competitive lives, knowledge was power—it still is, of course—and sharing personally valuable information was highly unlikely. Once our ancestors started cooperating, however, they would have been much more effective when everyone was on the same page.

The first step toward getting on the same page is shared attention. If I'm competing with other members of my group, I don't want them to know what I'm thinking, which means I don't want them to know where I'm looking, either. Whether I'm eyeing a potential mate or a tasty fig, I'll keep it secret so others don't get there first. But if I'm cooperating with other members of my group, then I will want them to know where I'm directing my attention. If a tasty prey animal comes along and I spot it first, I want others to notice it too, so we can work together to capture it.

Our chimp cousins are good at assessing visual perspective; they can discern what their fellow chimps are able to see from their vantage point. But chimps have evolved to make it more difficult for their peers to gather this information by hiding the direction of their gaze with brown sclera (the part of the eye that surrounds

* In his book *Social: Why Our Brains Are Wired to Connect*, Matthew Lieberman extends this argument to show the fundamental ways in which the human brain is wired to be a social instrument.

the cornea). If you look at a chimp's face, you can't really tell where it's looking without closely inspecting its eyes. In contrast, humans have evolved white sclera, which clearly advertise the direction of our attention. Following the gaze of a chimp, gorilla, or orangutan is no easy task, whereas the direction of our attention is readily available to others even when our faces and eyes point in a different direction.

The fact that we advertise the direction of our gaze in such a manner provides clear evidence that we typically gain more from others knowing what has grabbed our attention than we gain from keeping it a secret. Otherwise, our eye sclera would never have evolved away from those of the other apes. Some people have argued that such changes can occur because they benefit the group, and hence indirectly benefit the individual (as a member of the group). Such an argument is possible in principle, but the group benefit would need to be huge and the individual cost small for such a system to evolve. If the group benefits from knowledge that is costly to its individual members, in most circumstances the individuals still won't share the knowledge. Individual success determines what genes get passed to the next generation, even if individual success comes at a cost to the group.

As a consequence, when group goals conflict with individual goals, individual goals win out almost every time. Chimpanzees are far more self-oriented and far less group-oriented than we are, which is why they struggle to work effectively as a group. But once we moved to the savannah and found that cooperation was the key to success, we had the good fortune that group goals and individual goals aligned for the first time in the great apes. In other words, our expulsion from the forest created a new niche for apes who cooperated more than they competed. This evolutionary alignment of our group and individual goals eventually brought us to the top of the food chain, despite the conspicuous absence of any biological weaponry beyond our large brain.

In this sense, our cognitive evolution over the last six million years can be seen as a process of unintentionally pulling ourselves up by our own bootstraps. Our cooperative solution to a local climate crisis created, for the first time on this planet, a *social-cognitive niche*, and we spent the next few million years evolving new capacities to exploit this niche more effectively.

The Social Leap That Made Us Human

When our ancestors chanced upon a social solution to the challenges of life on the savannah they set in place a cascade of events that eventually led to our human origins, which is why I describe our move from the rainforest to the savannah as the "social leap." The leap from the trees to the grasslands was clearly metaphorical (and was really more of a shove than a leap, at any rate), but our leap to a social solution allowed us to move out of the shadows of the larger predators and set the stage for more complex social strategies.

Had we happened upon another solution to life on the grasslands (such as more effective burrowing, hiding, or running), I would not be writing this story and you would not be reading it. The choice our ancestors made was partially random but heavily constrained by the opportunities they had at their disposal.

Loss of our rainforest habitat could easily have been the end of us. If you replayed this vanishing rainforest scenario repeatedly, nine times out of ten I suspect we'd end up as timid versions of baboons at best, constantly looking over our shoulder for lions while keeping an eye on the nearest tree. Extinction or marginal existence was a much more likely outcome than our move to the top of the food chain. But some of our ancestors got lucky and found a solution to their existential crisis, and we are the beneficiaries of their resilience.

Our transition from the grasslands to Google was assuredly brutal and wildly inefficient, but that is the nature of evolution itself. Changes are continually wrought on this earth, and life either adapts or goes extinct. Indeed, humans would probably never have evolved at all if a massive asteroid had happened to miss planet Earth sixty-six million years ago. By smashing into the Gulf of Mexico and triggering global firestorms and climate change, that random bit of space junk eliminated all the enormous predators that had dominated our planet for over a hundred million years. We might have been able to drive off a lion or saber-toothed tiger by throwing stones, but our ancestors would have been tasty snacks for a *T. rex* no matter how many of us there were or how well we worked together. Our social leap was brilliant and seemingly prescient, but also highly dependent on a long series of events that just happened to go our way.

Most important of all, our social leap also transformed the evolutionary pressures on us. In response to the risks and opportunities inherent in our new life, we dramatically changed our mental proclivities and expanded our cognitive capacities over the next several million years. But that is the story for the next chapter.

2

Out of Africa

Three million years after leaving the forest, Lucy was walking upright, but she still looked far more like a chimp than a human. You probably wouldn't bat an eye if you saw her in the zoo, as there is little about her outward appearance that reveals her emerging humanness (see Figure 2.1). But Lucy understood that stones made useful tools, and some evidence suggests that she sharpened their edges to make them more effective. If true, this was a big step beyond chimps, who use stones as tools but have never been known to modify them.

A million years later, *Homo erectus* assuredly made tools of bones, sticks, and hides, but these have long since decomposed. The only tools we know they used when they first migrated to Europe and Asia were barely more sophisticated than the sharpened stones attributed to Lucy. These early stone tools *Homo erectus* adopted from their ancestor *Homo habilis* clearly

Figure 2.1. *Australopithecus afarensis*, aka Lucy. (Copyright © John Gurche)

Figure 2.2. *Homo erectus.* (Copyright © John Gurche)

made life easier—they were widely used and had remained unchanged for over a million years—but they were very simple indeed. I suspect that if you chanced upon one today, you could pick it up and skip it along the surface of a lake without ever knowing you'd held an object of great significance (see left side of Figure 2.3).

So how did *Homo erectus* successfully colonize Africa, Europe, and the southern half of Asia more than a million years ago with such a meager tool kit? No doubt their success stems from the fact that *Homo erectus* had a brain about two-thirds the size of our own. In contrast to Lucy, *Homo erectus* (Figure 2.2) would seem wildly out of place in a zoo as anything but a (rather rough-hewn) visitor.

The larger brain of *Homo erectus* enabled their most important tool: their enhanced capacity to work together. The fossilized remains of horses and elephants (often twice the size of modern elephants) butchered by groups of *Homo erectus* at multiple sites across Europe and Asia suggest that *Homo erectus* were not just living on the margins of their new world. Conceivably, *Homo erectus* were scavengers, nibbling on remains left behind by other predators, but the evidence suggests otherwise. Many of the marks that *Homo erectus*' stone tools left on the bones of these fossilized remains were made prior to the tooth marks of local predators. Additionally, the marks of *Homo erectus*' stone tools are often found high on the animals' leg bones, up near the torso, which is a part of the animal that predators consume first. If *Homo erectus* were scavenging, there would be little to no meat left on this part, and thus little reason for cut marks to appear there. These findings suggest that *Homo erectus* discovered how to kill some large and speedy animals with incredibly simple tools, an accomplishment that would have required extensive planning and coordination among groups of hunters.

There are several reasons to believe *Homo erectus* had the intellectual capacity to plan and coordinate hunts. First, *Homo erectus* went on to invent a much better set of stone tools than the Oldowan* tools they inherited from their comparatively dimwitted ancestor *Homo habilis* (left side of Figure 2.3). The Acheulian† tools created by *Homo erectus* were symmetrical and bifacial, which would have increased their utility and comfort of use (right side of Figure 2.3). In contrast to the stone tools of *Homo habilis*, if you chanced upon an Acheulian tool, I suspect you'd bring it home to show your friends.

Figure 2.3. An Oldowan tool on the left (Rosalia Gallotti) and an Acheulian tool on the right (Fernando Diez-Martin), both from 1.7 million years ago.

One way to study the demands involved in making these ancient tools is to make them ourselves, and numerous anthropologists have learned the art of stone knapping—that is, shaping tools from rocks by carefully whacking them together as our ancestors did. In one such study, modern stone knappers were placed in an fMRI magnet, which is similar to MRI machines used in hospitals, but with the added capacity of measuring brain activity along with brain structure. The knappers were shown partially completed Oldowan and Acheulian tools and asked to make decisions about what they would do next to finish them. By measuring brain activity while they made these decisions, the study allows

* The name Oldowan refers to the location of their initial discovery by Louis and Mary Leakey's team in the Olduvai Gorge, Tanzania.

† Named after the town of Saint-Acheul, France, where these tools were discovered a hundred years before the Leakeys' excavations in Tanzania.

us to see what sort of mental work is involved in making these tools. The brain scans showed that production of Acheulian tools demanded more processing from the front of the brain (the region involved in planning and coordinating activities) than did the production of Oldowan tools. This seems fairly intuitive when you look at the tools in Figure 2.3: it's clear that the Acheulian tools would require greater forethought than the simpler sharpening evident in Oldowan tools.

Not only did *Homo erectus* invent better tools, but the evidence suggests that they were the first species to plan for a world beyond their current needs. It seems odd, but monkeys and apes are incapable of planning for future needs they do not currently feel. For example, in one research lab, capuchin monkeys were put through a daily regimen in which they were fed only once per day. Capuchins can easily survive on such a diet, as long as the single meal is large enough, but they evolved to forage regularly, so they don't really like just one meal per day. Still, the monkeys never learned to save any of their extra food in anticipation of future hunger. Instead, once they had eaten their fill, they tended to fling their food at each other or outside their enclosure, rather than set it aside for a midnight snack.

Chimps are far smarter than monkeys, but they, too, appear incapable of planning for unfelt needs. By way of example, consider what happens when chimps fish from a termite mound. They first find a suitable bush, pull a branch off and strip away its leaves, and then carry this stick to a termite mound. They then plunge the stick into the holes burrowed in the termite mound and lick the swarming termites off it. This sequence of events shows that they can work through a series of stages to achieve their goal, but how thoughtful is that? Not terribly. As soon as they finish eating, they discard their termite fishing rod as if they'll never be hungry for termites again.

Similar results emerge in laboratory experiments, suggesting that chimps' inability to plan for unfelt needs manifests itself across numerous domains, even when advance planning would be incredibly useful. *Australopithecines* and *Homo habilis* were probably no more capable than chimps in this regard, as there is no evidence that they ever made their tools for use beyond the needs of the moment. In particular, there is no sign that Oldowan tools were ever carried far from the point at which they were quarried and made.*

In sharp contrast, *Homo erectus*' Acheulian tools have been found at great distances from where they were quarried and made. Thus, it's clear that *Homo erectus* perceived the future utility of their Acheulian tools even after they had finished using them to dismember a carcass. This is a major cognitive leap, but it makes sense if you consider the effort and intelligence necessary to craft Acheulian tools in the first place. The mental advances that enabled *Homo erectus* to make more complex tools also allowed them to predict that their needs would recur. For the first time in any species that has ever lived on planet Earth, we see evidence that our *Homo erectus* ancestors engaged in complex planning about the future, envisioning a world beyond their immediate needs.†

Finally, the most impressive evidence for *Homo erectus*' intelligence is that they invented division of labor. We saw clear hints of division of labor in the possibility that *Homo erectus* was successfully

* Of course, *Australopithecines* and *Homo habilis* didn't have clothes, much less pockets, so perhaps they found it easier to bang out new stone tools rather than schlep their old ones with them.

† Some readers might balk at this statement, rightly pointing out that squirrels, scrub jays, and many other species cache food and otherwise prepare for the future. In none of these animals, however, is there any evidence that they actually understand their future needs. Rather, they appear to have evolved caching as an instinctive behavior in the absence of foresight, without any understanding of why they behave as they do (in much the same way that female chimps evolved to leave their birth group to avoid inbreeding). These issues are discussed in detail in Thomas Suddendorf's excellent book, *The Gap*.

hunting large animals such as elephants and fast animals such as horses. When we look at their tool production sites, we see solid evidence that *Homo erectus* divided tasks among individuals to complete those tasks more effectively. For example, at a 1.2-million-year-old site in India, the manufacture of Acheulian tools was partitioned into different clusters, much as in a factory, with different aspects of production separated into different locations. The first step in the process of making these stone tools is to bash "flakes" loose from larger stones. Those flakes are subsequently shaped into different tools. At this Indian site, flakes were knocked loose in one location and retouched into their finished form elsewhere. If one person were making each tool from start to finish, there would be no reason to situate the production of different stages at different locations. Why carry a large stone ten meters away just to continue working on it? Rather, our ancestor would have sat down, knocked the flake loose, and then shaped it into a useful tool, all without moving around the site in any systematic fashion.

But different tasks require different abilities, so it makes sense to apportion the jobs among different people. Bashing the flakes loose required brute strength, as you have to give the larger stone a heck of a whack, in just the right spot, to dislodge a suitably shaped piece. This would have been a perfect job for the big, tough guys in the group, who could wield a large hammer stone and didn't mind if shards of rock went flying everywhere. In contrast, the finer knapping at the retouching stage required delicacy and coordination (much like embroidery), so women and smaller men would have been well suited for this job.

Numerous other examples suggest that our *Homo erectus* ancestors relied on teamwork to achieve their goals, but my favorite comes from an elephant butchery site on the Jordan River, just north of the Sea of Galilee. At this site, nine hand axes were found around the carcass of an elephant, and the elephant's skull appears to have been turned upside down with the use of a wooden branch as a lever, to allow the hunters access to the elephant's brains (a superb source of fat). Dislodging the skull from the spine and turning it over would have required several *Homo erectus* working together to control the substantial weight and awkward shape of the head. If the branch was in fact used as a lever, it's a virtual guarantee that some *Homo erectus* were pushing on the lever while others were balancing and rotating the head.

The chance of this endeavor working successfully in the absence of communication and coordination is near zero. Elsewhere in the site, there is evidence of nut cracking at one station, stone knapping at another, and processing of shellfish at still another. If these were modern remains, we might think they had set up food stalls, given how the tasks were so neatly divided into different locations.

Division of labor relies on the capacity to plan for the future, so it comes as no surprise that these two abilities would have emerged together. In combination, they greatly expanded what our ancestors were capable of achieving. These capacities moved us a good distance further down the social-cognitive pathway that *Australopithecines* put us on when they began to cooperate in their mutual defense.

Walking Upright

In chapter 1, I briefly mentioned that *Australopithecines* developed the body of a thrower as a by-product of bipedalism, or walking upright. But I failed to discuss why Lucy chose to walk upright in the first place. Now that we've discussed the different cognitive capacities of *Australopithecines* and *Homo erectus*, we can address this question, as it's one of the most important events in our evolutionary history. If we had not become bipedal, we almost assuredly would never have learned to throw so well, in which case the social-cognitive revolution that made us human might not have happened, either. But why walk upright? What did our ancestors gain when they stopped using their knuckles as a front pair of feet?

There are two answers to this question. First, we don't know.* Second, we have lots of ideas.† Some of our ideas have little or nothing to do with psychology. For example, long-distance walking appears to be more efficient for us on two legs than it is for apes on all four, and the disappearance of the forest would have put a premium on our ability to cover great distances. Part of the reason our ancestors started walking upright might have been simply to get from point A to point B without burning so many calories.

But some of our ideas are more psychological, in that they involve decisions about what to do with the hands. Perhaps our ancestors became bipedal to free up their hands to carry food, tools, or weapons. This proposal has been around for a long time, and it would certainly have given our ancestors an advantage to have extra food, tools, and weapons at their disposal.

As just discussed, the problem with this hypothesis is that there is no evidence prior to *Homo erectus* that our ancestors carried their stone tools any great distance, although they had been bipedal for a few million years. Our pre-*Homo* ancestors are unlikely to have carried food any great distance, either, for the same reason that they didn't carry their tools: they were unable to anticipate unfelt needs. If they were hungry at the moment, they would have eaten their food rather than carry it. If they were not hungry at the moment, they would have left their food behind or given it away rather than carry it (not realizing that they would get hungry again later). So, what would have motivated our *Australopithecine* ancestors to carry anything at all? Why would an animal that couldn't plan for unfelt needs ever choose to schlep anything across the savannah on a hot day?

* I admit, that's not much of an answer.

† Which is fair enough, because there was probably more than one cause of such an important event.

To answer this question, we need only consider what emotion would have been experienced by Lucy and her ancestors every time they headed out across the grasslands. What emotion would you feel if you had to cross the savannah on foot? I think the answer is fear. Every time Lucy and her ancestors set out across open fields, they would have been acutely aware of their vulnerability and scared of a possible attack by large cats or dogs. That fear would have motivated them to carry anything they could use to defend themselves—most likely a stick that could serve as a club or spear.

Carrying a club or spear is a lot easier when you can free your hands from walking, which would have motivated Lucy and her ancestors to walk upright. Even though Lucy couldn't plan for future needs, she could plan for current needs, and she would have felt the need for a weapon every time she set out across the savannah. Of course, we don't know if the perpetual desire for a weapon played any role in our transition to upright walking. Still, it is consistent with our cognitive abilities at the time, and it would have given our ancestors an advantage in a manner similar to the increased efficiency provided by bipedalism.

From *Homo Erectus* to *Homo Sapiens*

Division of labor created a new golden age for our ancestors, enabling the outcome of our joint activities to be much more than the sum of our individual efforts. With division of labor, groups had emergent properties that made them much more effective and deadlier than any groups had ever been before. More than four million years after our ancestors left the rainforest, *Homo erectus* put us back on a path to the top of the food chain by giving us something far more important than the safety of the trees. With division of labor, animals that were once our predators were now our prey.

As if division of labor were not enough, *Homo erectus* then sealed the deal with the single most important innovation in human history: the control of fire. Fire provided protection from the elements and predators, and also released nutrients and calories from food that are hard to extract when it's raw.* Compare the smell and taste of raw meat to a cooked steak or the palatability of a raw versus a cooked potato. In both cases, there is no comparison. One is nearly inedible; the other is delicious. With the control of fire, our ancestors changed their lives forever: no more returning to a cold, damp cave at the end of the day, no more night blindness, no more worrying about nocturnal predators while they slept, and no more living off the culinary equivalent of roadkill.

In his wonderful book *Catching Fire*, Richard Wrangham proposes that cooking played a critical role in enabling *Homo erectus* and later *Homo sapiens* to evolve such large brains. Despite having guts larger than ours, our fellow great apes cannot extract enough calories and nutrients from their diet of raw food to sustain a brain nearly as big as ours. Not only is our brain-to-gut ratio much greater than that of the other apes, but we also burn calories faster than they do. Cooking our food allowed us to evolve a faster metabolism to support such a large brain.

Cooking also allowed us to evolve greater fat storage, as a big brain is a risky drain on metabolic resources without a bit of blubber to serve as backup when times are lean. Finally, cooking freed our ancestors from the incessant chewing required by raw food. A typical chimp day involves about *eight hours* of chewing to soften food before it can be effectively digested. It's hard to imagine just how disruptive eight hours of chewing per day would be, particularly for a species like ours that is so dependent on spoken

* The control of fire also enabled most modern inventions, from the smelting of metals to the invention of steam engines and cars to rocket fuel—but, of course, those innovations didn't come along until much later.

language. Indeed, language might have evolved more slowly in a world of incessant chewing, due to the challenges of talking with one's mouth full.*

The control of fire illustrates one of the many ways our ancestors created the cognitive niche we occupy, and how culture and innovation can shape subsequent evolution. Once we learned to barbecue, we no longer needed such large teeth and jaw muscles to chew our food. This change allowed us to benefit from random mutations in genes that decreased the size of our jaw and teeth. These mutations would have been costly had they occurred earlier, as anyone with small or weakened jaws would have had greater difficulty subsisting on a diet of raw food. But once we learned to control fire, we moved further down the line from our large-jawed, small-brained ancestors to our small-jawed, large-brained selves. In this manner, our innovation played an important role in the subsequent shape of our head and facilitated our increasing emphasis of brain over brawn.

Our brains continued expanding as we evolved from *Homo erectus* to *Homo sapiens*, as did our capacities to plan and to understand each other. You and I are the outcome of this evolutionary process, as *Homo sapiens* emerged in Africa over two hundred thousand years ago. The complex cultures we developed with our large brains made us a success in every environment, and before long we had colonized all of Africa and begun looking beyond its borders. We see tentative signs of *Homo sapiens* in Arabia 125,000 years ago; and by 80,000 years ago we were heading out in earnest. By 65,000 years ago we'd made it to Australia; 45,000 years ago, we were pushing north into the Arctic; and 20,000 years ago, we reached

* Or perhaps not. Dentists somehow carry on conversations with patients whose mouths are full of instruments and cotton, and the ancient orator Demosthenes famously practiced speaking with a mouthful of pebbles. So perhaps we could have evolved to speak clearly while chewing at the same time.

the Americas. The Pacific Islands were the last places to be colonized, with New Zealand coming in last place, only 700 years ago.

Around the same time that we started colonizing the globe, there was an explosion of culture, art, and other evidence of symbolic thought. Although our species is more than 200,000 years old, the cultural richness commonly associated with *Homo sapiens* doesn't appear until less than 100,000 years ago. This does not mean there was a lack of symbolic thinking for more than 100,000 years, but more than likely it speaks to the minuscule chance that the products of our culture will remain intact for such long periods of time. Today we hang paintings in climate-controlled museums, but our ancestors just painted on cliff walls. Given their lack of concern with posterity, it's astonishing that we've found any art or artifacts from so long ago at all.

Whether complex culture emerged with the gradual onset of our species (or possibly even before us), or whether it took more than a hundred thousand years to accumulate, the eventual impact of our sociality was an extraordinary growth of knowledge and innovation. New tools, weapons, and art proliferated with the spread of *Homo sapiens*, and the key underlying changes that enabled this explosion were psychological. Long before the invention of writing (which is only about five thousand years old), human culture had become cumulative by virtue of our oral storytelling traditions.

Storytelling may have been another by-product of the control of fire, as the conversations of hunter-gatherers during the day differ notably from the stories they tell around the fire at night. During the day, people spend most of their time talking about ongoing social concerns and matters of economic necessity. But once night falls, communal fires are lit, and people gather in small groups, conversations blend into stories, and stories often reveal important lessons about how to live one's life and follow cultural rules.

Fire enabled us to extend our community time past daylight without the added risk of predation, and in so doing gave us a unique opportunity for socializing and reflecting, as the work of the day was no longer possible. The fact that hunter-gatherers use this time to pass on important cultural information raises the possibility that fire played a critical role in the creation of our knowledge base. Storytelling allows each generation to build on the information gathered by their ancestors, as cultures accumulate knowledge about how to deal with their local environment.

The importance of cumulative culture can be seen in almost every aspect of our lives, but one of the clearest examples can be found in the harrowing tales of early European explorers to the Arctic, Americas, Africa, Australia, and Asia. On countless occasions, intrepid and well-prepared adventurers perished or nearly died, while just around the corner, indigenous people who lacked their modern technology were well fed and sheltered. It was our capacity to learn from the experiences of others that gave *Homo sapiens* an enormous local advantage, with new strategies and innovations built on a platform of prior discoveries. As a consequence, each generation had no need to reinvent the wheel, and a child could acquire an understanding of the world that a few generations back would have been available only to geniuses. We see this effect today, with schoolchildren learning of the discoveries of Copernicus, Darwin, and Galileo, and there is no doubt that a somewhat slower version of this cumulative process has existed for over a hundred thousand years. No other animal can do this.

Complex Social Relationships Demand Large Brains

When you consider the vast body of knowledge required to survive in every climate on earth, it might seem that the social challenges our ancestors faced were trivial by comparison. The Inuit had to learn to hunt enormous creatures in treacherous seas and build temporary homes from ice to survive long trips on a featureless landscape. Sub-Saharan Africans had to learn to hunt monkeys using arrows tipped in poisons, such as those extracted from the *Strophanthus kombe* seed. And Aboriginal Australians had to learn to avoid deadly snakes and spiders while somehow finding food and water in one of the hottest and driest places on earth.

These challenges were extraordinarily difficult when humans first moved into these environments, but in the age before modern transport, humans moved slowly. Even our most nomadic ancestors spent most of their lives in familiar environments. The unchanging rules for dealing with the piece of the world they inhabited could easily be passed down to each subsequent generation around the campfire. As a result, cumulative culture and social learning ensured that the physical problems associated with predator, prey, and shelter weren't very cognitively challenging.

Unlike the physical world, however, the social world is dynamically interactive; my social strategies impact other people, who often change their behavior in response. Other people also hatch their own plans, and I need to figure out what's going on when I come home from a long hunt and everyone is whispering and giggling around the fire. Do I have mammoth fur stuck in my teeth? Was I cuckolded while I was out on the ice? Does that giggling even relate to me? Such complexities ensure that you cannot learn to deal with other people simply by following an unchanging set of rules that work well in the natural world, such as "avoid the spider with the red line on its back" or "find water in deep ravines." The shifting sands of the social world provide an ongoing challenge, as fellow humans change their behavior at their own whim, particularly if they think others are taking advantage of them.

All this might seem like little more than an intellectual game for our ancestors, but it's important to remember what life was like before the creation of government, law, police forces, and the various institutions of modern living that make our world safe and secure. As Steven Pinker notes in his superb book *The Better Angels of Our Nature*, life among hunter-gatherers was a dicey proposition, with murder rates that were typically much higher than those in the most dangerous cities of today. Being a hunter-gatherer might seem like an idyllic and carefree existence, but it was riskier than living in the scariest neighborhoods of the most dangerous cities today.

In the world of *Australopithecines*, *Homo erectus*, and eventually *Homo sapiens*, friends and neighbors did whatever they could get away with. Our closest analogue to their way of life is criminal gangs such as the Mafia and drug cartels, whose primary constraints are one another. Every morning when our ancestors got up from the cave floor, they joined a prehistoric episode of *The Sopranos*. If they wanted to survive until nightfall, they had to find a way to navigate the minefield of competing goals, complex and changing coalitions, jealous rivalries, and potentially dangerous outbursts from their peers.

Even being the strongest wasn't much of an insurance policy. Everyone sleeps eventually, and if others in your group decide you're more trouble than you're worth, any given night could be your last. Without law enforcement, everyone had to get by on his wits. The less savvy individuals were less likely to navigate this complex social world safely and less likely to convince someone to mate with them. As a consequence of these intense social pressures, we evolved a variety of new cognitive capabilities that were explicitly social in nature. The most important of these is Theory of Mind.

Theory of Mind

In their efforts to form and maintain alliances, create cooperative ventures, and just get through the day without getting whacked, our ancestors learned to anticipate one another's behavior. The best way to predict other people's behavior is to know their reasoning and goals, so we evolved Theory of Mind: the understanding that the minds of others differ from our own. Small children don't have this capacity, which is one reason why their sudden announcements and stories can be so difficult to follow—they don't realize that their listeners are often not thinking the same thing they are. But you can see the penny drop when it occurs to children that preferences and knowledge vary. Life suddenly becomes richer and easier when they realize that the world offers many mutually beneficial deals: I prefer red jelly beans, but you like the black ones; I'll play hide-and-seek if you'll play tag.*

Once our ancestors understood that others have different thoughts and feelings from their own, they began guessing what those thoughts and feelings might be. Other people's behavior provides the clearest clues to their thoughts and feelings, but behavior is most useful if you can discern the underlying intentions. Did she do that by accident, on purpose, or because she had no choice? This elementary form of mind reading is mission critical for understanding competing coalitions, particularly if the coalitions change in membership or goals across time.

It seems obvious that when someone stumbles and steps on our toes it was an accident, and when they walk over and stamp on our toes it was purposeful, but you know this only because your social information processor works so well. If you have pets, pay attention to the next time you accidentally hurt or scare them. My dogs become just as submissive and penitent when I accidentally step on their paws as when I scold them. With no capacity to differentiate intended from unintended actions, their ability to understand me and predict my future behavior is severely limited.

Along with these fundamental capacities of social perception, our highly interdependent lives ensured that we also evolved new

*The other great apes develop a partial version of Theory of Mind, but their partial version emerges only infrequently and manifests itself primarily in competitive circumstances. For example, alpha-male chimps will charge and separate other males who could threaten their position in the hierarchy if they appear to be establishing an alliance. There is no evidence, however, of chimps using even rudimentary Theory of Mind to work together to achieve mutually beneficial outcomes.

social emotions such as pride, guilt, and shame. These are often referred to as self-conscious emotions, as their development requires awareness of how others are appraising us, and they differ from other social emotions, such as anger and love, in that the focus is inward. These self-conscious emotions evolved to help us feel about ourselves as others feel toward us. They tell us immediately and forcefully which behaviors make us more valuable to our group and which behaviors devalue us.

We feel pride when we've done something that increases our value to our group, and the positive feelings associated with pride ensure that we seek out further such opportunities. We feel guilt when we've harmed someone in our group, and the negative self-directed feelings associated with guilt help us learn from the experience and avoid doing it again (before our friends give us the heave-ho). Shame is felt when we've done something to devalue ourselves in front of our group, and again the negative self-directed feelings ensure that we don't repeat the shameful behavior and experience further loss in status. Pride, guilt, and shame are critical parts of being human, and they help us function in the highly interdependent and cooperative groups that emerged with *Homo erectus* and that have made us so successful ever since.

The importance of these emotions can be seen in sharp relief when you consider the lives of people with a limited ability to experience them, such as sociopaths. Sociopaths are often charming and fun when you first meet them, but they struggle to get along with other people over any length of time because of their tendency to ruthlessly exploit others. If I experience no guilt or shame, there are no internal forces putting on the brakes when I see an opportunity to take what I want from you.

Taking rather than sharing or asking might seem like a successful strategy in the short term, but memories are long, and people who have been exploited or abused often recount this experience to others. Gossip plays a critical role in the spread of this type of social information, and soon even the most charming sociopaths find they are unwelcome in their community. In the hunter-gatherer groups of our ancestors, sociopaths struggled to find a community that would allow them to stay. As we will see in chapter 3, the creation of cities changed all that (and then social media changed it back again).

Theory of Mind for Teaching and Learning

Theory of Mind evolved to help us navigate our social world, but it has other advantages. Perhaps most notably, it dramatically increases our ability to teach (and learn from) others. If I have no idea what you're thinking, or that your knowledge differs from mine, it's hard for me to teach you. Where do I start? What do you already know, what do you need to know, and how can I best show you? But if I can discern the answers to these questions, I can intentionally use your knowledge base as a starting point for sharing new information. As a result, humans are incredibly effective teachers.

My favorite example of just how effective we are as teachers can be seen in the case of chimps learning to use stones to crack nuts. Chimps use simple tools, and in many parts of Africa, they have developed strategies for opening nuts by wielding stones as a hammer and anvil. When nuts are in season, mother chimps will often place some on a large stone and then use a smaller stone to crack them open. Their offspring typically sit nearby during this operation, and the mothers tolerate their children grabbing some of the fruits (or in this case, nuts) of their labor. The critical question for our purposes is how long it takes the offspring to learn this nut-cracking skill.

When I first heard about research on nut cracking, I guessed that young chimps would probably need about a year to learn the skill. After all, it's not easy manipulating irregularly shaped rocks and hard little nuts, and a few unintended finger smashings would dampen any student's enthusiasm. Additionally, chimps don't get the chance to practice very often, as nut cracking is a relatively rare event for them. So, it's not something they're going to learn overnight. You can imagine how stunned I was, then, to find that it takes chimps in the wild about *ten years* to learn to crack the various nuts they eat!

An animal trainer could teach a young chimp to crack a wide variety of nuts in less than a tenth of the time, but animal trainers have Theory of Mind, and chimp mothers do not. Chimp mothers do not know what their offspring do not know, and thus are limited in their capacity to teach them (and in their awareness that they *ought* to teach them). When they see their offspring make an error, they do seem to be able to correct it (such as when the young chimps hold the hammer stone wrong or place the nut poorly), but chimp mothers' lack of understanding of the error's source limits them to only infrequent and highly specific corrections. Small children are also poor teachers for the very same reason. Even when they have mastered a skill themselves, they don't know that others don't know it, and thus they struggle to teach their knowledge and skills to others.

Theory of Mind is a boon to learners as well. If I understand that another person has knowledge that I don't have, then I also understand that this person might impart that knowledge to me. This understanding prompts me to pay close attention to potential teachers, and to imitate their actions even if I don't discern their purpose. For example, if you are teaching me how to pitch a baseball, and you raise your front knee all the way to your chest before hurling the ball, then perhaps I should try to do the same. The motion looks ungainly and pointless, but you know better than I do.

This sort of imitation in the absence of understanding has clear links to Theory of Mind, and hence it's no surprise that it is uniquely human. In their classic experiment demonstrating this effect, Victoria Horner and Andrew Whiten of the University of

St. Andrews presented kids and chimps with a complex treasure box that had a treat inside. Horner and Whiten showed the kids or chimps how to open the box, but they made sure they included some irrelevant actions that played no actual role in opening it. For example, they first poked a stick into a hole in the top of the box, even though the only latch was on the side. When the external surface of the treasure box was opaque and observers couldn't see that the hole on the top had no function, the kids and the chimps both copied all the actions to open the box. When the treasure box was transparent, however, it was obvious which actions were relevant and which were not. In this case, the chimps modeled only the necessary actions and ignored the irrelevant ones, but the children continued to model the actions that were now obviously unnecessary.

This tendency has been labeled over-imitation, and it appears to be a universal human trait. It emerges among children in highly educated and industrialized societies, and also among children in small-scale societies in the Kalahari and remote regions of Australia who have little or no formal education. Over-imitation is an important human propensity, as it allows us to learn to do things even when we can't fully understand them. By assuming that our teacher knows best, we engage in the highest-fidelity copying we can, which enhances our effectiveness.

Over-imitation has great survival value, as it can facilitate the transmission of complex techniques that are often necessary in the preparation or detoxification of foods. Consider, for example, how people in the lowlands of Papua New Guinea have figured out how to eat the sago palm, which appears anything but edible. It turns out that the trunk of the tree contains a high concentration of starch, which can be harvested via a complex multistep procedure. After the tree is chopped down and the outer layer of the trunk peeled off, the inner part is pounded into sawdust. At this point the sawdust is entirely inedible, but it is then repeatedly rinsed with water. The warm tropical waters of New Guinea break down the starch molecules, causing them to separate from the wood of the tree and enabling them to pass through cloth filters.

The starchy water is then collected in large containers and left overnight so the starch can settle to the bottom. The water is then poured off the top, leaving a thick, starchy paste. This paste must be spread out and dried in the sun to prevent fermentation, which would otherwise make it toxic.* The subsequent flour is then stored in tubes made of sago palm fronds, where it can be kept for months and prepared in a variety of ways. No doubt a lot of trial and error went into the development of this process, but the beauty of over-imitation is that people don't need to know why they prepare the sago flour in this manner. They do it that way because they have observed others doing it that way.

Theory of Mind and Social Manipulation

It couldn't have been long from the origins of Theory of Mind to the day the first lie was told. I should clarify this statement by noting that deception existed long before lying. Deception is often practiced by plants and animals that pretend to be something they're not, such as insects that look like twigs and chameleons that change colors to match their background. These beings don't need to understand the mental states of others to achieve their deception.

Even complex acts of animal deception don't demand representation of the minds of others. For example, capuchin monkeys will occasionally make an alarm call in the absence of predators and then quickly eat the available food when the other monkeys scamper for the trees. Their use of this strategy is more likely when the food is concentrated in a nearby position, and hence more quickly consumed when the others run away. But even this rather complex

* Foods such as the sago palm make it abundantly clear how European explorers could perish in an environment in which the local population had plenty to eat.

strategy could be learned over time without any ability to know what the other monkeys are thinking.

In contrast to these types of deception, lying is a uniquely human form of social manipulation that requires substantially greater cognitive sophistication. To tell a lie is to intentionally plant a false belief in someone else's mind, which requires an awareness that the contents of other minds differ from one's own. Once I understand what you understand, I'm in a position to manipulate your understanding intentionally to include falsehoods that benefit me. That is the birth of lying.

Researchers have found that they can teach small children to lie simply by teaching them Theory of Mind. In the first test of this possibility, Xiao Pan Ding of Zhejiang Normal University and her colleagues brought three-year-old children into the lab who did not yet understand Theory of Mind, and either taught them Theory of Mind or an irrelevant task. In the Theory of Mind training, Ding and her colleagues showed the children containers, such as a pencil box, and then opened them to reveal that they held unexpected objects. They then asked the children what other people would think the box contained. Over time it began to dawn on the children that other people would tend to think as they had prior to seeing the unexpected contents. Children were trained across six different days on such tasks, with the goal of teaching them that other people don't necessarily know the information they have just learned.

After training in either the Theory of Mind or the irrelevant task, the children played a hide-and-seek game with the experimenter. In this game, the experimenter covered her eyes while the children hid a candy in one of two cups. When the experimenter opened her eyes, she asked, "Where did you hide the candy?" and then looked in whichever cup the child indicated. When the child told the truth, the experimenter picked up the candy and declared it her own. When the child lied by pointing to the wrong cup, the experimenter looked in the empty cup and then announced that she had lost and said that the candy now belonged to the child.

The critical question in this experiment was how often the children lied to keep the candy across this hide-and-seek game. Prior to the Theory of Mind training, the children never lied. After Theory of Mind training, the children lied an average of six out of the ten rounds. Knowledge is power, and this experiment demonstrates that Theory of Mind gives us the power of knowledge manipulation.

If you know small children, you can watch the development of lying in real time, because their early lies are so transparent. Children often begin lying to get out of trouble, such as when my three-year-old brother blamed our kitten when my mother asked him why he had splashed so much water out of his bath. This simple version of lying need not involve Theory of Mind,* but it does show that Theory of Mind is beginning to develop. From there it doesn't take long to start lying to gain all sorts of benefits that are difficult to come by honestly.†

I remember one afternoon in the park when a little boy joined my then-four-year-old son on the playground. After a few minutes on the monkey bars, the boy suddenly announced that he had a *Spider-Man* lunch box. My son hadn't seen *Spider-Man* before and must have been mystified about the nature of this apparently desirable object. Not to be outdone, however, he responded that he owned a "Leafman" and a "Grassman" lunch box. When the boy looked at me for confirmation, to see if my son had really bested him in both the lunch box and superhero departments, I did my best not to spill the beans by laughing. Although one naturally

* It's reminiscent of when Koko the gorilla used sign language to make the even more implausible claim that her kitten was the one who had ripped the sink off the wall.

† Children typically have a good understanding of Theory of Mind by the time they're four years old. Consider yourself warned.

tries to discourage lying in one's children, I was pleased to see the developing signs of sociality in my son's use of lying to maintain his status on the playground. Such uniquely human capabilities are products of our social nature and the critical importance of social success in the small groups in which we evolved.

Lying may be handy, but it is also a threat to relationships and to the social fabric of entire communities, as the advantages of our incredible communicative abilities emerge primarily when we tell the truth. People delight in the utility of their own lies but are furious when they catch others lying to them. Moral rules differ across cultures, but there are some universals, and one of the foremost rules in all cultures is not to lie. No one minds if you lie when you compliment their bad haircut, but self-serving and destructive lies are frowned on by every society on earth. When we see such universality in moral rules, we know that they combat a tendency in people to do otherwise, and serve an important human need. The pancultural condemnation of lying is clear evidence that all humans are tempted to lie, and that lying is a threat to group cohesiveness and coordination everywhere. Those facts, in turn, speak to the importance of cooperation and interdependence in human affairs.

If you track our cranial expansion across the six million years covered by the first two chapters of this book, you see that the story is quite extraordinary. A chimpanzee has a brain that weighs about 380 grams. Three million years of eking out a living on the savannah changed our bodies in important ways, but *Australopithecus afarensis*' 450-gram brain was barely larger than that of a chimp. Fast-forward another one-and-a-half million years to *Homo erectus*, and now our ancestors have a 960-gram brain, twice the size of that of *Australopithecus* (although they were a fair bit bigger as well, so the relative change wasn't as dramatic). Another million and a half years later, and *Homo sapiens* has an average brain weight of 1,350 grams. We added an entire chimp brain onto that of our *Homo erectus* ancestors. Why did the first three million years of evolution on the savannah give us a paltry 70 grams of brainpower, when the next three million years endowed us with almost a kilo?

The answer to this question lies in the fact that our expanding social capacities led us to evolve greater cognitive capacities to exploit new social opportunities. Without such a big brain, *Homo erectus* could never have controlled fire, and without controlling fire, *Homo erectus* could never have evolved an even larger brain. More important, without the social complexity of *Homo erectus*' lifestyle, there would have been little reason to spend the metabolic energy necessary to support such a large brain in the first place.

The accelerating brainpower that emerged over the last six million years was both cause and consequence of the social changes experienced by our ancestors. We created the social-cognitive niche when we learned to cooperate on the savannah for mutual defense. Over the next several million years we continually invented new ways to leverage and exploit our expanding social-cognitive abilities.* We remained hunter-gatherers for the six million years that elapsed from when we left the trees until very recently, but our place in this world changed dramatically. Cooperation and division of labor expanded our capabilities, transitioning us from prey to top predator.

* At the time of this writing, a pair of papers has just been published highlighting the possible role of NOTCH2NL genes in human brain expansion. These genes appeared in our line three to four million years ago, just as our ancestors' brains started to grow in leaps and bounds. Prior to the creation of our social-cognitive niche, these genes would have cost more than they were worth, and thus would not have proliferated. But after we started cooperating, the gains in social effectiveness enabled by greater intelligence easily offset the increased metabolic costs brought about by these brain-expanding genes.

3

Crops, Cities, and Kings

HOW AGRICULTURE PUT THE FINISHING TOUCHES ON OUR PSYCHOLOGY

Agriculture emerged about twelve thousand years ago in the Middle East, soon thereafter in China and the Americas, and over the next few thousand years in many other places. For ten to twenty thousand years prior to the advent of agriculture, people in Europe, the Middle East, and China gathered wild cereals and ground them to make flour. To accommodate their increasing reliance on plant foods and associated agricultural implements, our hunter-gatherer ancestors slowly shifted away from their nomadic lifestyle. This transition involved numerous changes to their way of life: a house instead of a tent, clay pottery instead of gourds, and stoneware such as the mortar and pestle, which was handy for grinding wheat but a nuisance for nomads.

Once our ancestors started farming, they didn't immediately put aside their spears and bows; hunting continued alongside farming just as it did alongside gathering (and just as it does in many farming communities today). From our vantage point, the invention of agriculture was a watershed event, but it probably wasn't such a big deal to our great-great-. . .-grandmother when she decided to plant some of the seeds she had gathered. Rather, it likely struck her as a convenience to know where her preferred plant would grow the following season, and seemed worth a try. Indeed, I suspect she knew not the seed she'd sown.

Although farming provided our ancestors with somewhat greater food predictability and stability, numerous costs emerged in the transition from a nomadic hunter-gatherer lifestyle to the stationary life of a farmer. Let's start with the one that would bother me the most: lack of proper plumbing. Twelve thousand years ago everyone defecated outdoors, but hunter-gatherers had the advantage of moving on before they suffered the consequences. Farmers weren't going anywhere, so over time they completely fouled their drinking water with their own feces. This process of fecal poisoning was disastrous for their health.

We have a modern analogue to this situation in the high rate of outdoor defecation that exists in some parts of India, where it is a major cause of gastrointestinal illness and malnourished children. The situation faced by early farmers was equally dire, as their sedentary lifestyle exposed them to this new disease vector. Indeed, some scientists believe that agriculture led us to evolve a tolerance for alcohol rather than a distaste for it, because alcohol killed many of the bacteria that farmers were unintentionally introducing into their own drinking water. Beer was safer than water. In addition to the diseases farmers caught from their own feces, the animals they kept also proved to be a major source of illness, as human epidemics often have their origins in domesticated animals (e.g., swine and avian flu).

Moving on from the toileting and illness situation, the next major problem faced by farmers concerned the quality of their agricultural diet. Although most people in modern industrialized societies access a wide variety of foods year-round, this is unprecedented. Our hunter-gatherer ancestors generally achieved a well-balanced diet, but they achieved it over long periods of time. When berries, fruits, or nuts were in season, they gorged themselves on those, and moved on when they had fully exploited the local resources. Early farmers, in contrast, had far more restricted diets, with less seasonal variety and starchier foods from the cereals they grew. As a consequence, farming reduced the nutritional quality and variety of our ancestors' food.

In the process, farming also radically changed the balance of our various oral bacteria, with the unfortunate result that nastier bacteria flourished in our newly sugar-laden mouths. Despite never owning a toothbrush or floss, hunter-gatherers rarely got cavities or gum disease. In contrast, the teeth of early farmers were typically half-rotten, and by medieval times they were utterly foul. Their high-starch, low-variety diet not only resulted in poor oral health, but also led to a decrease in their stature and shorter life spans than were enjoyed by earlier hunter-gatherers. Indeed, it is only in the last few generations that we have surpassed the height of hunter-gatherers, and it was only with the advent of modern medicine that we began to live longer (excluding violent deaths).

Finally, farming was a seasonal catastrophe when it came to working hours. Hunter-gatherers in "immediate-return societies" (meaning those who eat today what they catch today) typically spend about six hours per day hunting, gathering, preparing meals, and mending tools. The rest of their time is spent socializing and relaxing until it gets dark, at which point storytelling and dancing by the firelight are common activities. While it is true that traditional farmers often work only an hour or two each day during quiet times, during the busy planting and harvest seasons they spend every daylight hour working and relax only after dark. Depending on the availability of water, the number of planting seasons per year, and various other factors, traditional farmers may or may not work more hours than hunter-gatherers, but they certainly work much harder for sustained periods of time.

When we weigh up the costs and benefits, we see that farming afforded our ancestors some assurances against starvation, but at the cost of various new illnesses, reduced stature and longevity, excruciating halitosis, and often a far longer working day. The end result was that early farmers worked harder to achieve a worse life than their ancestors had.*

Farming may have been a disaster for individual farmers, but it was a success story at the population level: it allowed large numbers of people to live on land that would have supported only a small group of hunter-gatherers, and it increased people's reproductive rate. This increase in reproduction and population density caused farming groups to outgrow and eventually displace hunter-gatherer groups. The numerous migrations of farmers across Europe and Asia and back into Africa over the last ten thousand years are a testament to the fact that farming communities can dominate hunter-gatherer communities, even though any individual hunter-gatherer was typically a healthier specimen than any individual farmer.

Domination by farming didn't take place overnight, and weather patterns, drought, and other disasters often favored hunter-gatherers, who could more easily move on when their immediate environment became uninhabitable. As a result, farmers and hunter-gatherers coexisted in Europe, often cheek by jowl, for at least two thousand years.

* You may ask yourself if the dawn of the digital age has resulted in a similarly lousy deal for us. Until recently, phones were attached to landlines, and if you wanted to write someone, you had to send a letter through the mail. In those days, I received a few calls a day from my friends—no one bothered calling when people were likely to be out—and a letter or two every day. Mobile phones, the Internet, and email have enabled us to be in touch all the time, but that connectivity comes with a cost. I love that I can contact my family, friends, colleagues, and students so easily, but instant global communication also means that work requests pop up on my phone or laptop a hundred times a day. Like the first farmers, I had no idea what was in store for me when I signed up for an email account thirty years ago. I was pretty sure that it would make my life easier, and although I'm not so sure now, it's hard to opt out.

The Psychology of a Farmer

Farming requires mental capacities that first emerged in *Homo erectus*: division of labor, effortful preparation of tools, and planning for the future. But the psychological changes required to shift from a hunter-gatherer to a farmer demanded more than just these capabilities, which had been in place for many millennia prior to agriculture. Farming also required a change in attitudes and values to match new demands and opportunities. Consider the lifestyle of a tropical hunter-gatherer versus that of a farmer:

Tropical hunter-gatherers typically live in immediate-return societies. Because it is nearly impossible to store meat in the tropics, and because even the best hunters often come home empty-handed, immediate-return hunter-gatherers share everything they catch with the rest of their group. This practice of universal sharing creates an insurance policy that serves everyone's interest by smoothing out the rough patches that might otherwise lead to lean times or starvation.

Tropical hunter-gatherers follow game movements and opportunities for gathering plant foods (e.g., ripening berries and fruits), so they own no more than they can carry. By virtue of their nomadic lifestyle, hunter-gatherer societies are formed of interlocking groups that split apart and re-form in new ways whenever people decide to break camp and try their luck elsewhere. If you don't like someone in your group, it's pretty easy to decide to go west with your family when that person decides to go east with his. Everyone connects with members of their overarching group when they re-form into new camps, and hence they spend their lives surrounded by people they know well, but the subgroups that form any individual camp are fluid.

In contrast to hunter-gatherers, who live each day as it comes, farmers focus on tomorrow. Their labor centers primarily on preparation for the harvest, which is a hugely important and high-effort event. Farmers often own a large number of agricultural implements, as farming is greatly facilitated by devices that enable them to prepare the fields, bring in the crops, and turn the crops into edible products (such as the grindstones that long preceded agriculture itself). Farmers are stationary for the obvious reason that you can't bring land with you, and once you've acquired the tools, cleared the land, and planted your crops, there is a substantial cost in leaving it behind. If you don't like someone in your farming community, chances are pretty good that neither of you is going anywhere.

Although farmers around the world have differing rules about sharing the fruits of their labor, it is very rare for crops to be shared beyond family and those responsible for helping with the harvest. Karl Marx suggested that people should produce according to their abilities but share according to their needs, and this maxim describes hunter-gatherers reasonably well. But the history of communism suggests that agricultural people don't share very well outside their family.

The problem with the Marxist utopia is that people can free-ride on the efforts of others. If you are required to share with me, then I'm tempted to put in a little less effort because I know your hard work will leave me well fed. Once you see me slacking off, you don't want to be a sucker, so you slack off a bit yourself, and pretty soon everyone is barely working. The free-rider problem is a vicious circle and can quickly destroy a productive community if there is no way to police everyone's contribution.

Recall that our ancestors first encountered this free-rider problem when their fellow *Australopithecines* ran away from predators rather than contribute to their collective stoning. Such free riding was easily witnessed, and threatened ostracism or punishment brought slackers into line. Our *Homo erectus* and *sapiens* hunter-gatherer ancestors also solved the free-rider problem relatively easily, because daily catches became daily dinners, and it's pretty easy to see who's contributing and who isn't. Hunters who never brought home the bacon often found themselves in a group of one if they couldn't find a way to make themselves useful. But farming is a long game, and it's not immediately obvious on any given day how hard everyone else is working. When I end up with a smaller crop at the end of the season, maybe I was a slacker and am undeserving of your generosity, but maybe my soil is poor or my plot was attacked by pests.

This problem of detecting free-riding farmers was compounded by the fact that farming inevitably led to an increase in the size of communities. Because hunter-gatherers could move house in a matter of hours, and because they shifted their communities numerous times each year, people spent their entire lives in small groups of family and friends (or at least nonenemies). Whenever groups got too large and people started to bicker, hunter-gatherers broke into smaller groups that went their own ways.

Because some types of land are more suitable for farming than others, farming communities grow in areas where the land is particularly productive. Within just a few generations, agricultural communities grew much larger than the hunter-gatherer communities of their ancestors. This increase in size meant that, although early farming communities were small by today's standards and connected by a network of interdependent relationships, people no longer knew everyone well in their community. In such large groups it is difficult to discern who works hard and who is a slacker, so people simply stopped sharing as much.

These problems among farmers were magnified by the fact that traditional sharing goes beyond the end result to include the means of production. Hunter-gatherers share not only meat, but also weapons and tools. If they own more than one knife, bow, or gourd, their friends and family will often ask for it. This approach to life is unsuitable for farmers, as minimum levels of livestock, land, and equipment are necessary to make farming viable.

This incompatibility can be seen today among communities that are moving from hunting and gathering to a market economy. For example, some members of !Kung San hunter-gatherer groups have become involved in trade, herding, and day labor with nearby farming communities, and are often paid in livestock. The sensible long-term strategy for such individuals is to breed their new animals, using them for milk or eggs, but they immediately face demands from family and friends to share their newfound wealth. If they refuse these requests they are branded as stingy, which is a social disaster. If they acquiesce, their hard-earned gains disappear in front of their eyes. The possession of livestock often makes their lives worse rather than better, because private ownership is inconsistent with their communal culture.

Ultimately, these demands of farming shifted our psychology from one of communal sharing to one of private property. Such a shift probably didn't require new psychological adaptations, but it did demand something of a cultural upheaval. I witnessed the challenges of this shift myself when I worked with a remote Aboriginal community in northern Australia. The manager in charge of one of the environmental monitoring and clean-up teams was impressed by their productivity and hard work, and at the end of their first contract he offered them a raise. He was mystified by their indifference to his offer, and the fact that some of them declined it outright.

When he probed for the underlying reasons, he discovered that when members of the team go home to their extended families at the end of the workweek, everyone asks them for whatever money they've earned. There was no benefit to getting a raise; if anything, it was frustrating to watch even more money disappear. The manager solved this particular problem by offering high-quality meals onsite rather than a raise, and the employees were delighted with the opportunity to benefit from their hard work without being seen as stingy by their family and community.

Private Property

Private property has a lot of pluses, but in a world in which people achieve their own outcomes, inherent differences in ability, effort, and opportunity eventually result in inequality. Some people are smart, talented, hardworking, or lucky, or choose their parents wisely, and they end up with a lot of stuff. Other people, not so much. This fact is blazingly obvious in today's world, but it required a seismic shift in our hunter-gatherer psychology to accommodate this emerging reality.

Adapting to inequality was one of the most difficult challenges our hunter-gatherer ancestors faced as they transitioned to life on the farm, but it was necessary. Inequality inevitably follows from the demands and opportunities of ownership and stockpiling associated with agriculture. In today's world, we see the first signs of inequality when we move from immediate-return hunter-gatherers, who rarely have a paramount leader, to hunter-gatherers who also tend their own gardens (known as hunter-horticulturalists) and often have hereditary chiefs. Similarly, in the ancient world we see the roots of inequality among those hunter-gatherers who started to make the transition to farming. Even before our ancestors domesticated plants, some had large houses while others had small ones. Some people were buried in elaborate garments and jewelry while others were placed into the ground unadorned.

The emergence of inequality in some places but not others led scientists to wonder how and why inequality arises. It turns out that the literature in ecology and biology, particularly research on animal territories, is helpful for addressing these questions. Territories are the animal version of inequality. Among territorial animals, males can't attract a mate unless they have a territory, and the better their territory, the more mating opportunities they have. But only some animal species maintain territories. Some animals have territories they vigorously defend from other members of their species (and sometimes other species, as in Disney's *The Lion King*, with the ongoing conflict between lions and hyenas), and some animals don't.

Biologists have found that the best predictors of whether a territory will be defended are its resource density and reliability. Only concentrated and predictable resources provide enough benefits to offset the costs of a territory's defense. Because grass is so common, of low yield, and widely spread, it's typically not worth the effort for herbivores to defend their bit of savannah from other herbivores.* In contrast, the herbivores themselves represent high-density food packages, and hence lions will defend their patch of savannah from other lions.

Given this sensible decision making on the part of our furry, feathered, and scaly cousins, it's no surprise that humans follow the lead of other animals when deciding whether to accrue and defend resources. Immediate-return hunter-gatherers rarely attempt to defend territory, as the resources are often unpredictable and low in density. When some hunting grounds are better than others, hunter-gatherer groups do come into conflict with other groups over these locations. Even in such cases, however, one person or even one family could never defend such a large and low-density resource from another family. For this reason, variability in hunting grounds leads to inequality between groups, but not within them.

In contrast to life in the tropics, some hunter-gatherers live in ecologies that enable food storage and also have highly dense, predictable resources. For example, Native Americans who fished for salmon in the Pacific Northwest caught far more than they could

*There are, of course, various exceptions to this rule. For example, at certain times of year, male wildebeest defend their patch of grass from other male wildebeest (but not from other herbivores such as zebra and impala) and use their territory to entice females.

eat during seasonal salmon runs, so they dried their catch for later consumption over the winter. This sort of ecology promoted the development of inequality, as families attempted to dominate and defend the best fishing spots during the salmon run. Such families incorporated others by offering some of the fish in exchange for help with the harvest, defense, preparation, and storage of the catch. Arrangements such as these were slow to emerge, but they eventually led to social customs that shifted from strict egalitarianism to institutionalized inequality.

Similar arrangements cropped up every time resources became predictable and sufficiently dense. For example, hunter-horticulturalists rarely defend the wild plants they gather if intruders enter the local forests, but they always defend the crops they plant in their own gardens. By the time societies move to full-scale agriculture, norms develop around the notion of property ownership and rights. These norms are so important that even modern societies fail to function properly when there is poor rule of law and people cannot count on the state to defend their property rights. Under such circumstances, people are unwilling to invest the necessary effort to maintain and improve their property, as they have no confidence that their efforts will be rewarded over the long term. In contrast, when people live in a society that protects the rights of property ownership, there is clear utility in investing in their own future by improving their equipment, lands, and homes.

In the case of our early agricultural forebears, the ten- to twenty-thousand-year period of pre-agriculture and comparatively stationary living likely played an important role in laying the groundwork for the psychological changes that support private property and inequality. Discomfort with inequality, conflicts about property rights, and many other societal norms probably shifted slowly over generations as people began to transition from communally sharing nomads with virtually no possessions to stationary farmers who owned the sources of production.

Private Property and Gender Inequality

With the advent of private property, the rules for provisioning the household changed dramatically. Hunter-gatherers require the combined efforts of Mom and Dad to feed the children. Dads typically hunted for large game and the higher-calorie but lower-probability proteins and fats, and moms typically gathered the lower-calorie and higher-probability plant foods. Because the product of the hunt was shared by the whole community, most people's needs were met most of the time. The wives and children of the best hunters were not much better off than the wives and children of the worst hunters, although they had the comfort of knowing that whatever group they were in, there would be at least one good hunter.

Once people began to accumulate private property, those who had stuff were in a much better position to provision a family than those who did not. Wealth and ownership were easily discerned by others, and these assets would rapidly have become attractive in a potential partner. Perhaps not surprisingly, the impact of wealth on reproductive success is clearer for men than it is for women. Particularly in communities that allow polygyny, but also in communities where men have informal mistresses (i.e., everywhere), wealthy men have the potential to have many more children than poor men. Hunter-gatherers in immediate-return societies simply cannot support more than one family of children, but private property enables wealthy men to support huge numbers of children.

Numerous princes and kings throughout history have had hundreds of children, with the rapacious Genghis Khan quite possibly holding the world record (with 8 percent of Asia potentially his descendants).* Women's reproductive potential, in contrast, is not influenced by the number of partners they can attract; a man with twenty wives can have two hundred children, but a woman with twenty husbands cannot. Wealth was important to women because it helped their children survive, but beyond the moderate wealth required for survival, additional resources do not enable women to have more children (although great wealth does give women more grandchildren, through the increased capacity of their sons to attract more partners).

Due to these sex differences, wealth was less important for reproductive success for women than it was for men, so men were typically much more motivated than women in their pursuit of wealth. All else being equal, mothers and fathers were also more motivated to pass their wealth on to their sons than to their daughters, again because the reproductive benefits that accrue to wealthy sons are greater than the benefits that accrue to wealthy daughters. Partially because of these sex differences, gender inequality spread across the globe along with private property.

Men typically have more power than women in immediate-return hunter-gatherer societies, but women are more equal to men in such societies than they are in other types of foraging or agricultural societies. Agriculture also disrupted gender equity by virtue of the activities involved in preparing the fields and harvesting them. Plow-based agriculture in particular demands greater musculature to work the fields, and hence created a sex-based division

* Not only did Genghis Khan rape his way across Asia during his conquest, but he and his Khan descendants also had enormous numbers of concubines. These practices suggested that his genes should be overrepresented in Mongolia and the surrounding areas. Geneticists eventually found an unusual gene on the Y chromosome that was common in the area of Mongolian conquest but rare elsewhere, and that could be traced back to approximately one thousand years ago. In combination with the historical record, these data suggest that this gene proliferated from Genghis himself (who lived eight hundred years ago).

of labor, with men working the fields and women preparing the food inside the home.

Consistent with the cultural "stickiness" of such habits, societies that practiced plow-based agriculture continue to have less female participation in the workforce outside the home than societies that practiced other forms of agriculture (e.g., hoe-based agriculture, in which women can effectively work the fields alongside men) or no agriculture at all (e.g., in which men hunt and women gather). In this way, agriculture destroyed the imperfect (yet pervasive) gender equity that existed in immediate-return hunter-gatherer societies.

Agriculture Created Government, but Also Hierarchy, Exploitation, and Enslavement

Once we decided that private property and inequality were acceptable, we started down a slippery slope to all sorts of misery. If it's okay for me to own more stuff than you do, then it follows that it's also okay for me to survive and even thrive while you starve. After all, if I share with you today when I can easily afford to do so, I might find myself in a difficult situation tomorrow because my generosity has left me without sufficient reserves. Our hunter-gatherer psychology did not support this logic given that today was paramount and tomorrow was a bridge to be crossed later, but this logic fits well in an agricultural psychology that emphasizes ownership and long-term plans. Nonetheless, watching fellow group members suffer from a position of relative comfort was inconsistent with numerous psychological changes brought about by our increased cooperativeness and interdependence over the last six million years. As a consequence, new patterns of thinking were required to mitigate the dissonance that this disparity aroused.

First and foremost in these psychological changes was the idea that some humans are better and/or more deserving than others. Although hunter-gatherers vary in traits such as strength and cunning, they tend to be fiercely egalitarian and strongly resent anyone who acts superior. Indeed, the most successful hunters tend to self-deprecate the most, so that others don't envy their apparent superiority or get the sense they are lording it over them. As Christopher Boehm describes in his fascinating book *Hierarchy in the Forest*, this psychology leads to interesting patterns of communication—the bigger the catch, the more the hunter downplays it, along the lines of the following:

SID: How'd you do today?
RICHARD: Not much luck, although perhaps I could use a little help dragging the tiny carcass back home.
SID: *Hmm, sounds like he got something big.* . . . Should I get a few others to come along, or can you and I carry it together?
RICHARD: I'd hate to have you bug anyone else; it's so little and sickly that we'll probably leave it there anyway.
SID: *Holy crap, he's killed a giraffe! Time to call in the whole camp* . . .

As our ancestors moved away from their hunter-gatherer lifestyle, the talented and hardworking accumulated nicer homes and more stuff. From there, people naturally concluded that property ownership is earned and that better people have better stuff. The passage of time and intergenerational inheritance would have eroded the strength of such a relationship, as lazy and dimwitted people could benefit from the generosity of their parents, but that problem was solved psychologically by our deciding that some *bloodlines* are better than others (a concept that is still evident in our fascination with erstwhile royalty).

Thus, the first psychological step away from our hunter-gatherer lifestyle was the willingness to agree that some people are better than others and that inequality is an acceptable outcome. Once you accept that "fact of nature," why stop at rich and poor? Why not king and peasants, or master and slaves? Of course, that's exactly what happened. Worldwide, egalitarianism was kicked to the curb, and newfound beliefs in innate superiority led to all sorts of human suffering. We know little of how this process initially unfolded, but records of European imperialism allow us to track the underlying psychology via parliamentary debates, newspapers, and other sources of public discussion as inequality spread to new people and places.

Such records show that opportunities for European exploitation and enslavement were rapidly followed by justifications based on differential qualities of "us" and "them," or even differential humanness. Rudyard Kipling's "White Man's Burden" and the Manifest Destiny of the American settlers are both examples of how colonialists were able to recast their exploitation or slaughter of indigenous populations in positive and even moralistic terms. Of course, not everyone bought into this line of reasoning, and these arguments were often the topic of fierce debate. However, history shows that those in favor of exploitation and slaughter almost always carried the day, at least if we judge colonial powers by their actions and not their words.

Just as illuminating, exploration and exploitation of Africa and the Americas reveal European efforts to force inequality upon local populations. Such enforcement proved to be much easier when the locals were already agriculturalists, and accustomed to the psychology of inequality, than when they were hunter-gatherers and found inequality highly aversive. When Europeans attempted to conquer agricultural communities, such communities often quickly acquiesced to their new masters. The peasants in such societies were already heavily taxed and poorly treated by their own aristocracy; what difference did it make to them who took the reins? In contrast, hunter-gatherer societies typically fought against the colonialists, sometimes for generations, before finally being conquered by the larger and better-armed European forces.

In their illuminating book *Why Nations Fail*, Daron Acemoglu and James A. Robinson provide numerous historical examples of this effect. For instance, when the Spanish began their conquest of South America, one of their earliest settlements was at the site of modern Buenos Aires. The settlement was a colonial failure and soon abandoned because the local hunter-gatherers refused to work for the Spanish, even under extreme duress. When the Spanish ventured farther inland and encountered agriculturalists in Paraguay, they easily subjugated the local people by conquering and replacing the aristocracy, while maintaining their system of forced labor.

Agriculture and inequality demanded complex societal organization, as people who could store food learned to band together against outside invaders and thieves within their own community, creating rules of ownership and commerce. Such rules could have been applied equally, but early governments were typically power struggles among competing factions that resulted in fiefdoms designed for exploitation. Only a few governments prior to the American Revolution were intended to serve their constituents. The Enlightenment and rediscovery of individual rights (a concept that is strong among hunter-gatherers) clearly played a major role in the improvement of government, but one need only look at the flawed state of most democracies to see the temptation on the part of the elite to exploit their position for personal advantage.

Chapter 7 provides a more thorough discussion of this topic, but for now the point is that the psychology that enables a state to shift from an oligarchy to a representative democracy is complicated. Perhaps the single most important factor is heterogeneity of interests. When the elite all benefit from the same set of rules (e.g., high import taxes to prevent foreign competition), there is a real risk that the interests of the elite will take precedence over the interests of the masses. But when society is highly heterogeneous and the interests of the elite contradict one another (as occurs when some are agriculturalists, some are manufacturers, some are merchants, and some provide services), the most likely compromise is a system of rules that is basically fair.

We see this same psychology in the emergence of preferences for fairness among children. When children are very young, they prefer to receive more than others. As they enter middle childhood, they start to realize that today's advantage could easily become tomorrow's disadvantage. Because other people outnumber the self, it doesn't take children long to realize that unfairness is more likely to favor others than it is to favor the self, so they start to prefer outcomes that are relatively even between self and other. The safest psychological bet is equitable distribution, as this serves the self most reliably in the long run.

To this day, my two children believe that the older one cares more about fairness than the younger one, but in reality, he just figured out the odds before she did. When my daughter was still asking to draw another card in Candy Land or Monopoly, my son was old enough to realize that this approach was a bad idea, and he protested strongly. What seemed rather remarkable was that he was equally adamant when we offered to bend the rules in his favor. I was foolish enough to take pride in his strong sense of justice, but in fact, experience had taught him that fairness was in his own interest in the long run.

When people feel that they cannot dominate their group and guarantee long-term advantage for themselves, the same psychology emerges, and preference for fair rules predominates. We may not realize that self-interest is the source of our preference for fairness, but Lord Acton's famous aphorism that "power tends to corrupt, and absolute power corrupts absolutely" is testament to the fact that in our hearts we know that no one can be trusted.

From Villages to Cities

Homo erectus invented division of labor and benefited greatly from it. But the true potential of division of labor can be realized only when people become specialists, spending a lifetime pursuing a particular interest or talent. Such specialization was impossible until large enough groups crowded into small enough spaces that people could focus on their own interests without worrying that the crops wouldn't be brought in or the laundry wouldn't get done. Cities made that possible. If you live in a small village, maybe no one is interested in being a blacksmith, or maybe the only blacksmith falls in love with a girl from another village and leaves, so you'd better be able to shoe your own horse. But if you live in a city, numbers alone guarantee there will always be someone who can generate the goods and services you need yet are unable to produce for yourself.

Michelangelo could never have existed in a world of small villages; nor could Newton or Shakespeare. Although my favorite barista wouldn't call herself an artist (a point on which I disagree), her glorious caffè macchiato could never have existed in a world of villages, either. In short, without cities, people couldn't afford to focus their energies on a single skill to the exclusion of all others, so almost no one could develop the expertise to create something new and wonderful.*

With the advent of cities and true expertise, the world finally moved from the low-yield or even zero-sum game† it had always

* I should note that some societies created highly specific divisions of labor that enabled people to become specialists in the absence of cities, and even the absence of agriculture. Although such cultural practices create true expertise, they provide very little choice and variety with regard to areas of specialization. In such societies, you cannot simply choose to become an artist or a mason because that is your passion. Rather, there must be a need for the role, and cultural rules must indicate that you are suitable for it. Additionally, expertise in one's chosen domain demanded the development of cities, but you no longer need to live in a city to develop expertise. Once modern transportation became available, people everywhere had the opportunity to pursue their interests.

† A zero-sum game is a contest or negotiation with a fixed reward, such that one person's gains always mean another person's losses. Sharing a pie is a zero-sum game; when I take a larger slice, you're left with a smaller one.

been, in which fortunes could be made only at the cost of others' misfortunes. Expertise allowed the overall size of the pie to grow, as new things could be created that benefited everyone. Uruk, in eastern Iraq, might be the very spot where, six thousand years ago, the pie started expanding. Rome wasn't built in a day, and neither was Uruk, but by five thousand years ago, it housed tens of thousands of people, and writing, pottery, and trade goods from ancient Uruk spread across the Middle East. Once we had Uruk, and all the other cities that followed, the world started to become a much richer and better place.

Change rarely comes without cost, and the move to cities was no exception. For the first time in history, our ancestors faced a life filled with acquaintances and strangers. This might not seem like a big deal if you live in a city and see strangers every day, but it's bizarre for those who aren't accustomed to it. I remember experiencing my own version of this discombobulation when I visited Shanghai about twenty years ago. There was a well-known market I wanted to see, and when I got there, I discovered that it was basically a human sardine can. So many people were crowded into such a small space that everyone was physically pressed on by everyone else. I thought it looked like a market mosh pit, and an experience I should sample, so I put my arms to my sides and waded in. Within just a few minutes I felt as if I were going to have a psychotic break, as strangers' faces, armpits, and various other body parts pressed on me from all sides, and there was nothing I could do by way of escape. I wormed my way out of there as best I could, and my memories of Shanghai have never really recovered from that experience. For hunter-gatherers and early farmers, the crush of strangers they encountered in the first cities must have seemed equally disconcerting.

In small villages, everyone knows everyone else, at least by reputation. People know who can be trusted and who can't, who has a hot temper and who is easygoing, who is competent and who is hopeless. Cities are full of strangers, so city-dwellers do not know what to expect when they encounter each other in a shop or a bar. We have responded to this development with two important changes in our psychology.

First, the safest strategy when dealing with unknown strangers is politeness, and indeed cultures that have high levels of violence often have high levels of politeness. For example, the American South is famous for its politeness and friendliness. But Southerners are also much more likely to react with violence, particularly when they have been treated dishonorably. Greater politeness and friendliness may seem inconsistent with greater violence, but in fact they are opposite sides of the same coin. When you're among people who take offense violently, the most sensible strategy is to be polite to everyone, particularly people you don't know.

With the move to cities, we shifted to a strategy of automatic politeness to strangers. Among hunter-gatherers these conventions don't exist, as people respond to each other in keeping with their relationships. But among strangers in cities, we expect a certain level of decorum and respect on their part, and in turn we offer the same. If we slip and fall, we expect strangers to walk around rather than over us, and not to laugh at our clumsiness. If we drop our groceries on the ground, we expect strangers to help us pick them up, or at least not grab them and run away.

Cities have also caused us to rely much more heavily on outward appearances. The aphorism "don't judge a book by its cover" would never have been used in hunter-gatherer societies, because there was no reason to do so. Everyone you interacted with was known to you.* Why judge a book by its cover when you've read every page? In contrast, we know very little of the capabilities and proclivities

* People encountered strangers when they came across unfamiliar groups, but such circumstances were fraught with danger and did not involve the sort of casual mingling that strangers engage in today.

of the people we encounter in cities, so we rely much more heavily on outward displays.

For example, talking a big game is more impressive from a stranger, whose stories might be true, than a close associate whom you know to be exaggerating. Similarly, overconfidence on the part of a stranger is typically interpreted as a sign of competence, even though we might roll our eyes when our friends and neighbors show inflated self-views. As is discussed more thoroughly in chapter 7, this excessive reliance on outward appearances created by city living can be problematic, simply because we can't calibrate people's claims against their known behaviors.

This is not to say that we are incapable of judging people by their appearance. To the contrary, we are surprisingly good at it. But "surprisingly good" just means that we are a fair bit better than chance at judging who is friendly, who is competent, and who is malevolent. Lots of unfriendly people seem friendly when you first meet them, lots of incompetent people maintain an air of competence, and a fair number of sociopaths wheedle their way into our hearts and wallets (and even high office) before we discover their true nature.

Fortunately, there is a lot of seemingly trivial information we can use to judge others accurately. For example, a glance at people's music playlists, college dorm rooms, or apartments provides reliable information about their personalities. Personal appearances are equally informative: your clothes, hairstyle, and many other aspects of your personal grooming reveal your personality and temperament.

In my own case, prior to my first date with my wife (and hence prior to her consoling me about being bested by the prepubescent twerp at the Ohio State Fair), I was already thinking that maybe she and I were a match. That thought didn't occur to me with regard to the other women I was seeing around that time, and I suspect that there was something about my future wife's appearance or way of speaking that communicated to me that we were fundamentally compatible. Research shows that such thin slices of behavior can be surprisingly revealing, and also that seemingly irrelevant factors such as scent can be very important, although we know little about how these processes work.

How the Internet Brought Us Full Circle

Perhaps the most significant cost that has emerged in our move to cities is the greater scope they provide for the dark side of human nature to predominate. In a small community there are no strangers, so it's hard to lie, cheat, steal, or otherwise take advantage of others. When our hunter-gatherer ancestors misbehaved, there was really no way to escape the consequences: gossip ensured that their reputation would catch up with them. In a city, by contrast, it's easy to exploit friendly and trusting people and then move on before your duplicity is discovered. Modern levels of residential and occupational mobility allow sociopaths to outpace gossip their entire lives.

This challenge is now being met head-on by social media, which is returning us to the close-knit, intertwined lives of our ancestors. Many websites are explicitly designed to communicate reputation. TripAdvisor not only lets me know whether a hotelier maintains a clean establishment but also gives me the opportunity to air my grievances or delight after my visit. Such websites give otherwise powerless consumers a great deal more ammunition in their fight with Goliath corporations.

Other websites have proliferated to provide private individuals access to the reputations of strangers. Searching someone's police record is a clear example, but Uber, Airbnb, and eBay are all business models that rely on the reciprocal transmission of reputation. If I don't know you, I might be nervous about letting you ride in my car or stay in my house. But if I know how you have been rated by lots of other people who let you ride in their cars and stay in their houses, I have a pretty good sense of what sort of person you are.

By transmitting reputation, these websites minimize the risk of exploitation. Because every interaction is rated by both parties, buyer and seller are both motivated to deal with each other fairly and honestly. I might be tempted to treat your car or house like a trash can, but I know I'll pay a steep price for that behavior the next time I try to call a car or book a vacation rental. Just as our ancestors were inclined to deal with each other fairly and honestly because the social costs were too high if they did otherwise, the social (and hence financial) costs are too great for users of these platforms to cheat one another.

Along with these formalized methods for sharing reputations, platforms such as Facebook have also allowed private citizens to let the world know when they have been exploited. Consider the Australian con artist Brett Joseph, who repeatedly charmed his way into women's hearts with the goal of accessing their wallets. Although he successfully ripped off one of his first victims, she was determined to prevent it from happening to other women, so she set up a webpage that provided his photo, name, and modus operandi. Despite numerous attempts to repeat his swindles, judicious googling by his potential victims stymied him time and again. He even moved to the United States in an effort to escape his reputation, only to be caught out when he applied for a marriage license and admitted to his fiancée that he hadn't been using his real name. That admission struck her as odd, and it didn't take long before she discovered his true identity. As was the case with our hunter-gatherer ancestors, Brett Joseph's reputation now precedes him no matter where he goes.

In this sense, social media and all the reputation buttons we have at our disposal are an important weapon against the sociopaths, slobs, cheats, and freeloaders who would otherwise take advantage of the kindness of strangers. As is nearly always the case, however, these benefits come with costs. The most notable cost is the manner in which social media can crush people through relentless attacks by those who don't know the perpetrator and hence aren't inclined to temper their reactions with sympathy. Overreactions to bad behavior are a repeated occurrence, with people's lives disrupted for what were objectively minor infractions. People tweet stupid and insensitive jokes to their friends, or make what they thought were private comments, and within hours they find themselves unemployed.

Consider the story of Walter Palmer, the Minneapolis dentist who shot Cecil the lion in Zimbabwe. Unbeknownst to Palmer, Cecil was a beloved lion who lived in Hwange National Park and was being observed by Oxford University's Wildlife Conservation Research Unit. Palmer had paid $54,000 for the opportunity to bow-hunt a lion and must have been thrilled when such a magnificent animal wandered off the National Park and onto the farm where he was hunting. He failed to kill Cecil with his bow and arrow, but eleven hours later his team tracked down the wounded lion and shot him. It was only then that they found the monitoring device around the animal's neck.

Some good came of this tragedy: many countries banned the import of trophy lions; more than forty airlines announced they would no longer carry them; lion conservation rules were strengthened in several African countries; and donations increased to wildlife conservation organizations. In contrast, the reactions to Palmer himself were vicious. Many people are appalled by trophy hunting, but Palmer had broken no laws and had not intentionally killed a lion under observation. His actions were really no different from those of thousands of other hunters who legally kill animals for the thrill of it. Nonetheless, he was met with international condemnation, his dental practice was protested, his property was vandalized, and he received death threats from animal rights activists when he returned home, prompting him and his wife to go into hiding and hire personal security. One year later, there were still attacks against him in the media, and some news outlets posted photos of him and his whereabouts.

Many people probably think Palmer got what he deserves, and maybe they're right. But my guess is that if they knew him personally, their anger and disgust might be moderated by other information about him. Through the internet we have regained the ancestral benefits of reputation broadcasting, but we pay an enormous cost in the overreactions of strangers who know the perpetrators only through a singular misdeed and are unaware of their positive qualities. Social media has given us some of the benefits of the small communities in which we evolved, but it may prove to be the case that ancestral living was a package deal, and the benefits of modern forms of reputation broadcasting are outweighed by the overreactions of strangers.

Our journey from the East African rainforests of seven million years ago to the cities of today has been extraordinary, as we survived and thrived in environments that could easily have killed us off many times over. Our proclivities and capabilities evolved throughout this period, as we changed slowly from chimplike beings to the people we are today. But surviving and thriving are not the whole story, as evolution depends primarily on reproduction. We turn now to the last part of our prehistory: the impact of our ancestor's mating habits on our modern psychology.

4

Sexual Selection and Social Comparison

Imagine a world in which people who wanted to save for retirement were required to find someone of the opposite sex who would agree to partner with them to create a joint account. Imagine further that both parties got to draw equally from their interest-bearing account when they retired, but the rules dictated that for every dollar the man deposited into the account, the woman had to deposit a million. Finally, imagine that people could set up as many joint accounts as they desired, so long as they could find a willing partner of the opposite sex. This is an outrageous set of rules, but if such a world existed, what would you do?

Your answer likely depends on whether you're male or female. If you're a man, you would probably be happy to establish retirement accounts with any woman with a pulse. With that level of return on your investment, how could you lose? No matter how reprehensible she is, it's a pretty sweet deal. In contrast, if you're a woman, you're in a tight spot. On the one hand, you need to create a joint account so that you can retire with some sort of savings. On the other hand, the rules of the account are stacked against you so strongly that you're probably going to be very picky when you choose a joint account holder. He's occasionally grumpy? Forget it. No reason to deal with a difficult joint account holder when there are plenty of nice people out there. He sings off-key? Not worth the pain and suffering. A more tuneful or quieter person will assuredly come along who'll want to join you. And the list goes on . . .

As is evident in this example, whoever invests more resources into a joint outcome gets to call the shots when it comes to choosing a partner. The sex that contributes a smaller amount will compete for access to the investment made by the sex that contributes more, with the result that the high-investing sex will be much choosier than the low-investing sex. Robert Trivers had this insight in his famous paper on "Parental Investment," and it turns out that it explains sex differences in mating strategies and mate competition across the animal kingdom.

In biology, the animal that produces the larger gamete (i.e., sex cell or, in this case, egg) is the female, and the one who produces the smaller gamete (in this case, sperm) is the male. For many animals, producing gametes is the sum total of their parenting effort. The female frog lays her eggs, the male frog sprays sperm across them, and then the two usually hop off into the sunset, leaving their hundreds or thousands of fertilized eggs to hatch into tadpoles and eventually grow up to be big and strong, or perhaps to be eaten by the next fish to swim by. There is no parental effort in such species beyond the act of producing and depositing eggs and sperm, yet even that degree of parenting is a significant investment.

Because the volume of biological material necessary to produce eggs is much more than that required to produce sperm, Trivers' theory suggests that male frogs should compete to be chosen and female frogs should do the choosing. And that is exactly what happens. For example, in many species of frogs, the males sit around a pond or suitable site and croak as long and loudly as they can. The females hop back and forth to compare the volume, pitch, and duration of the croaks, and when it becomes clear who the best croakers are, the females mate with the winners.

Frogs work hard to catch enough flies to produce their eggs and sperm, but humans put frogs to shame with the amount of parental effort we devote to our offspring. It's difficult to quantify the emotional and social energy that we put into our children, but we can calculate the biological demands. Because we're mammals, the obligatory female investment in reproduction is substantially greater than just producing the larger sex cell. After producing the egg, human females must gestate it for nine months, with every gram of nutrition given to the fetus pipelined from the mother.

After giving birth, our female ancestors typically breast-fed their babies for another two years, with almost every calorie received by the baby being sourced from the mother. This might not seem like a big deal in a world in which the next meal is only as far away as the pantry—indeed, my wife delighted in the fact that our infant son was a human liposuction machine—but our ancestors found the caloric demand of creating and raising a baby extremely burdensome. Remember, they had to hunt down and kill or dig up and drag home every piece of food they ate.

As a result of this biologically mandated difference in parental investment, for every unit of energy that a human male must donate to produce a child, human females must donate much more than a million times as many.* This is where the joint retirement account analogy comes into play: women have more skin in the game, with the result that men tend to compete for women and women tend to be much choosier than men.

One important way that men compete for women is by showing

* I haven't actually done the math to compare the metabolic energy required to produce one dose of sperm versus nine months of gestation and two years of lactation; I'm just choosing a large number. I suspect the truth is more one-sided than a million to one by several orders of magnitude.

that they will provide food, shelter, and protection for them and their offspring. Given the caloric cost of having a child, this is a major concern for women, and men have evolved to respond to this concern by showing evidence that they are good providers. Our ancestors displayed their ability to provide by being good hunters, and modern men show their ability to provide by going to college, getting good jobs, or displaying wealth. As I discuss in chapter 10, because humans form long-term, exclusive bonds in which they work together to raise their children, human mate choice is also a mutual decision. For this reason, in addition to these sex differences, there is a fair bit of competition among both men and women to get the most desirable mate.

Sexual Selection

Reproduction is the currency of evolution. If every animal had the same number of surviving offspring, there would be no evolution. Survival is important, but only in terms of living long enough to reproduce and pass your genes on to the next generation. Organisms that successfully raise lots of offspring and that facilitate the reproduction of their close relatives, pass along their proclivities into the next generation. Organisms that fail to do these things represent the end of the line for their genes, and their particular proclivities disappear from the gene pool. In this manner, traits and behaviors that are associated with reproductive success become more common than traits and behaviors that are not.

Due to this process of selection, we evolved to enjoy activities that enhance our reproductive success and dislike activities that don't. For example, almost all adult humans enjoy sex and almost all humans find feces distasteful. It's no surprise that having sex increases our chance of passing our genes on to the next generation, and also no surprise that eating feces decreases those chances. It's important to note, however, that having sex and eating feces are not really parallel activities with regard to our reproductive success. Having sex has little bearing on survival but directly relates to creating offspring. In contrast, eating feces diminishes our odds of survival and thereby diminishes our chance of reproduction.*

This distinction between factors that directly impact survival and factors that directly impact reproduction was crucial for Charles Darwin, the father of evolutionary theory. If I live a thousand years but don't reproduce, my incredible longevity is evolutionarily irrelevant. But if I live long enough to see my children through to adulthood, my survival facilitated my reproductive success. Of more direct impact is whether I can attract a mate and successfully reproduce, and hence the capacity to attract and retain a mate is of central importance in evolutionary theory. Darwin used the term *sexual selection* to describe the evolution of those processes that enhance our access to members of the opposite sex.

Sexual selection is a powerful force in evolution. If there is a trait that members of the opposite sex find distasteful, that trait will tend to disappear in the population (even if it facilitates survival) because owners of that trait will have trouble finding mates. For example, it's conceivable that being a super-timid guy and hiding at the first sign of danger might have facilitated survival, but most women would have found this an undesirable trait because such a partner probably wouldn't protect her and her children. As a consequence, there haven't been too many super-timid guys, or at least not many who were willing to act in a super-timid way when women were around. Similarly, if there is a trait that members of the opposite sex find attractive, even if that trait diminishes our chances of survival, it can become common in the population because owners of that trait will have more mating opportunities.

This observation leads us to ask why members of the opposite sex would ever find survival-diminishing traits attractive. By way

* Eating feces probably also diminishes our chance of getting a date.

of answering this question, consider the difference between males and females of one of the most remarkable birds on the planet: peacocks and peahens. A peahen is a sensible bird, with mostly gray-brown feathers and a tail that is just a little longer than it needs to be. If you were a hungry tiger, you could walk right by a nesting peahen and never know it was there, as it would blend into its surroundings. In contrast, a peacock is an audaciously bright bird that sports the most outrageous tail in the animal kingdom. The peacock's bright coloration would attract the attention of even a myopic tiger, and its cumbersome and heavy tail would make it much easier to catch once spotted.

Indeed, the bright and heavy peacock tail seemed like such an impediment to survival that Darwin famously wrote to the Harvard botanist Asa Gray, "The sight of a feather in a peacock's tail, whenever I gaze at it, makes me sick!" Darwin is one of history's greatest scientists because he had incredible insight, but also because he worked so hard to address the weaknesses in his theory. In the case of the peacock, he eventually realized that the costs of the bright color and huge tail in terms of survival were offset by the gains in reproduction. To understand why the peacock's tail facilitates reproduction, we need to consider why certain traits are attractive.

What Does It Mean to Be Sexy?

Every organism manages the best it can, by whatever means necessary, which ensures that deception is rife among all living things. If deception helps an animal avoid predators or get a mate, then you can count on its being deceptive. My favorite deceptive mating strategies can be found in the species of birds and fish in which some of the males pretend to be females. In these species, males who aren't large enough to fight off other males have evolved a strategy of feigning femininity, which allows them to consort with females without being challenged by bigger males. These small males then seize the opportunity to sneak an occasional copulation, thereby passing on their genes even though they would never win a fight with the larger males who are interested in the same females.

Perhaps the most extraordinary example of this strategy was witnessed among cuttlefish by Culum Brown of Macquarie University and his colleagues. Cuttlefish can change their color at will, and when they aren't disguising themselves to match their background, females often show one color pattern while males show another. Brown was observing cuttlefish in their tanks one day when he noticed a male displaying male coloration on the half of his body facing the females, and female coloration on the half of his body facing the males. This deception allowed the two-faced male to court the females while simultaneously preventing the other males from attacking him.

Such deceptions show how animals can be dishonest to achieve their goals. No surprise here; evolution is amoral, and animals adopt whatever strategy works. But life is a coevolutionary struggle, with every organism evolving counterstrategies to deal with the strategies of their competitors. In this sense, evolution is an endless arms race, with animals developing new ways of responding whenever they are faced with novel capabilities on the part of their prey, predators, or parasites. If the balance tips too far in favor of one or the other, then extinction becomes a distinct possibility. When prey animals evolve a disguise, predators evolve greater detection abilities. If males evolve ways to deceive females into choosing them when they are a poor choice, females evolve ways of seeing through this ruse. Females who are easily duped by low-quality males are less likely to leave successful offspring in subsequent generations, so female preferences are shaped by evolution to favor males who show honest signals of quality.

An honest signal of quality is one that is impossible, or at least very difficult, to fake. For example, I can tell you about my yacht and brag about my ski vacations, but unless you've seen my balance sheets, you don't know whether I'm loaded or a liar. Talk is cheap. But if I fly you to St. Moritz and scoot us around in a Maserati, that's a much better signal that I'm well-heeled (and generous, too). I can also casually mention that I graduated first in my class at Harvard Law School, but again, I could be lying to impress you, so you'd be wise to doubt my claim. In contrast, it's more convincing if you watch me easily solve a Rubik's cube, as that allows you to see my brain in action. It's more impressive still if I can solve a Rubik's cube without looking back at it once I begin, because I'm demonstrating that I can win with a handicap.

Due to the importance of honest signals, humans are very adept at detecting even subtle cues that indicate quality. I remember one spring morning in college when I was sitting in the courtyard waiting for my friends to join me for breakfast. I happened to have a copy of the *New York Times* and a pen, so I decided to attempt the crossword puzzle while I waited (despite being hopeless at crosswords). As I sat staring at the first clue, pondering "Brought to tears, possibly," an older alumnus and his family walked by. He saw me working on the puzzle and sat down to join in. I could have used his help, but before he got a chance, his wife grabbed him by the ascot and pointed out, "He's working on the Sunday puzzle with a *pen*, dear. He does not need your help." She interpreted my subsequent claims to the contrary as pure politeness, and they walked around the corner to admire the blooming trees.

A few minutes passed by and, as luck would have it, my brilliant friend Katrin plopped down next to me as I continued to wonder who was brought to tears, and why only possibly. She solved the first few dozen clues as fast as I could write the answers, at which point I realized I was now merely a scribe and shooed her away so I could give it a try. With perfect timing, the alumnus and his family passed by again, and when he tried to sit down for a second time, his wife admonished him, "Dear, he's solved nearly half the puzzle in less than five minutes. He does *not* need your help." Trying to solve a crossword puzzle with a pen is not really an honest signal, but solving it is,* and that's why we pay attention to such details.

To return to the peacock, its bright color and extraordinary tail are honest signals of male quality because they are an *enormous* handicap, and that's why the peahen finds the bright colors and huge tail so alluring. Growing a large and brightly colored tail is a bit like trying to solve the Sunday crossword with a pen. Any idiot can start the process, but it takes a pretty special organism to carry it off. Peahens are responsive to this fact, so they are attracted to peacocks who are able to survive year after year despite schlepping around five feet of brightly colored tail feathers.

A mature peacock could well be the most honest signaler in the animal kingdom, but lots of birds adopt a toned-down version of this strategy. Numerous species have long tail feathers, and when biologists artificially shorten or lengthen their tails, they find that the females respond accordingly, flocking to the long-tailed lads and shunning those who've had their tail feathers clipped. Not only do females of many bird species find long tails sexy, but as we saw with peacocks, color is also important. Bright colors advertise your presence to predators, so female birds can use the brightness of a male's colors to deduce his quality. A brightly colored bird must be strong and fit to survive looking that way, whereas a dull-colored lad may well be slow and clumsy, and thus would be less suitable as a mate.

Keep in mind that female birds aren't thinking this through. Rather, those females who prefer brightly colored males leave more offspring in the next generation, and hence pass on the tendency to prefer bright colors. The same holds true for human preferences—we don't need to understand why we find an hourglass shape attrac-

* In this particular case, it was an honest signal of *Katrin's* intelligence.

tive in women or a triangular shape attractive in men. Evolution simply ensures that most people have these preferences.

So, birds are evolutionarily primed to prefer bright colors, but not all colors are created equal. Animal immune systems rely on plant pigments called carotenoids (such as beta carotene) to function properly, but animals are incapable of producing carotenoids on their own—they rely on plant photosynthesis to do the job for them. Animals who are struggling to fight off infections need to dedicate every bit of carotene to their immune systems, whereas healthy animals have an excess of carotenoids in their system. The reds, oranges, and yellows that many birds display on their feathers are made up of carotenoids. Because birds can afford those colors only if their immune system is robust, bright colors (particularly bright reds, oranges, and yellows) are honest signals of internal quality, even though they are being worn on the outside.

Bright reds are inherently honest signals of quality due to the metabolic cost in producing them, but there are other honest signals of quality that anyone can produce. For example, some birds have evolved to signal their dominance through the size of a black or brown patch on their chest or throat. Black feathers are no more costly than white feathers to produce, yet in these species, a larger dark patch indicates a higher position in the hierarchy. This was confusing when it was first discovered, as even the weakest males can easily grow a larger patch,* and hence it seemed that female birds of these species were falling for cheap talk.

But further study revealed that other males take these color patches very seriously, and don't take kindly to subordinates strutting around with large dark patches as if they're dominant. Wearing

* It might seem odd to suggest that a bird can "decide" whether to grow a large or small patch, but dominance is associated with hormonal changes, which in turn lead to bodily changes. These effects can be seen in numerous species, such as when a cichlid fish develops bright coloration to signal that it now holds a territory.

the wrong-size patch is an affront to every other male in the flock, and when biologists tested this possibility by painting larger black patches on low-status males, they essentially painted a target on their chests. Every other male in the flock will go out of its way to attack such usurpers, in an effort to show them that they're just posers and would be wise to shrink their patch. In such cases the size of a patch becomes an honest signal, even though it can be produced without biological cost, thanks to the social consequences of wearing a larger patch than one should.*

What are honest signals of quality in human males? Size is a good indicator, as you can't grow to be six feet tall if you're not well nourished and healthy. Muscles are also a good indicator, as is athleticism, for the same reason. But, of course, humans care a great deal about brains as well as brawn, and markers of a good mind are also honest signals of quality. That is why women tend to find a sense of humor sexy. Not only is it fun to be around funny people, but a good sense of humor requires an agile mind to draw connections that other people find funny.

Facial symmetry is also an honest indicator of quality, as humans have evolved to be symmetric, but illnesses and accidents can disrupt that symmetry. Symmetry is a signal of health and genetic robustness (or at the very least a blessed life), and thus women tend to find it attractive. If you google "Brad Pitt" and look at his face, you'll be struck by its symmetry. If you google "Lyle Lovett," you'll be struck by the indicators of a difficult life.

Symmetry, strength, height, and humor are also honest signals of quality in women, but they are not the primary markers of quality that attract men's attention. Because being fertile is much more biologically demanding for women than it is for men, men are more interested in women's honest signals of fertility. Most notably, men

* Growing a larger patch than you've earned is a lot like wearing a Red Sox cap to Stan's Sports Bar in the Bronx during a Yankees game—cheap to buy, expensive to wear.

are attracted to signs of youthfulness and an hourglass shape—both signals of female fertility. If humans were like chimps in their reproductive habits, and women became more successful mothers as they aged, then no doubt men would be strongly attracted to older women. The evolution of menopause among human females has changed that equation (more on this issue in chapter 10), with the result that men are more attracted to women during their most fertile period, from their late teens to mid-thirties.*

The Theory of (Social) Relativity

Sexual selection and mate competition are the driving forces behind the power of *relativity*, that is, the importance of our relative standing compared to others. For example, women evolved to prefer men who are kind, generous, funny, cute, and smart; but even if I'm none of the above, they'll still choose me if I'm their least-awful option. It doesn't really matter how smart or attractive I am, so long as I'm smarter and more attractive than the other available men. Similarly, it won't do me a bit of good to look like Henry Cavill and have a brain like Albert Einstein if everyone in my group is better looking and smarter. Although the chances of that are slim, the point is that our absolute levels of any given trait don't matter that much; what matters is how we stack up against the other relevant members of our group. For this reason, people engage in social comparison all the time.

We begin this process of social comparison initially to learn about ourselves and our social standing. Am I strong or weak? Fast

* Although there is a lot less work on partner preferences and sexual attitudes among members of the LGBTQ community, the research that does exist often paints a similar picture. For example, my colleagues and I found that whether men were straight, bisexual, or gay, they tended to regret missed sexual opportunities more than they regretted bad sex they'd had. In contrast, whether women were straight, bisexual, or lesbians, they tended to regret bad sex they'd had more than missed sexual opportunities.

or slow? Rich or poor? Despite what we might think, there are no absolute answers to these questions. It all depends on how we stack up against other people. If I can bench-press more weight than the people who are close to me, I'm strong. If not, I'm weak. Our nearest neighbors are also our most important competition, so we tend to look to those closest to us to answer these questions.

The problem with this tendency to look locally is that it can lead to a very skewed worldview. I remember sitting with my friends late in our senior year of high school, reflecting on the highs and lows of the last four years. One of my friends said that her biggest regret was her lack of athletic success. She was the state champion in at least two sports, competed on the varsity team in several others, and the hardware hanging off her letter jacket would have put a Russian general to shame. I asked her how on earth she could feel that way, given that she was one of the best athletes I knew. She replied that she had failed in her dream to make the Olympic team. That aspiration may sound absurd, but two of her siblings had competed in the Olympics, so that was her standard of achievement.

We can see the effect of relativity in countless domains, and sometimes it makes perfect sense to worry about everyone else. For example, imagine I invented a pill that increased your intelligence by 50 percent, and I offered you one. Immediately after taking the pill, you'd feel so much smarter—all sorts of problems that had seemed complex before would now be child's play. Particle physics and calculus problems would be amusing ways to entertain yourself while waiting for a haircut. But now imagine that I offered everyone else *two* pills. You'd be sitting there waiting for your haircut, and everyone around you would be discussing ideas that were way beyond your capacity. In a matter of moments, you'd go from feeling like a genius to feeling like a fool.

In this example, it doesn't matter how many pills you take if everyone around you is always one pill ahead. Even if you can solve differential equations in your head, if other people are smarter, you'll get the least interesting jobs, your friends will think you're slow, and more generally you'll tend to be left behind. Relativity matters in these instances. But because of the importance of sexual selection, we also get caught up in relativity when other people don't really matter.

Consider Ilyana Kuziemko of Princeton University and her colleagues' work on "last-place aversion." They found that the greatest resistance to raising the minimum wage was among people who were making slightly more than that. Despite the fact that raising the minimum wage has a high probability of helping individuals who are making just above minimum wage at some point in the future, their concerns about their relative standing outweighed the potential advantages that might come from a higher minimum wage. Such reactions might seem like cutting off your nose to spite your face, but sexual selection provides the underlying logic.

Concern with relative standing has also been found in our primate cousins, and is most famously demonstrated in capuchin monkeys by Sarah Brosnan and Frans de Waal of Emory University. In their experiment, they trained monkeys to return a pebble to the experimenter after it was placed in their cage, and the monkeys were paid in cucumber slices for their efforts. The monkeys clearly regarded this payment as fair, given that they learned and sustained the appropriate behavior based on these cucumber rewards.

The critical phase of the experiment took place when the monkeys witnessed another monkey receiving a grape (a preferred food) for the same action that they were completing for a cucumber slice. If fairness is an absolute judgment, it shouldn't matter what reward the other monkey receives for doing the same task. If a cucumber slice was good enough a moment ago, it should be good enough now. On the other hand, if fairness is a *relative* judgment, then it matters a great deal what the other monkey is paid.

Consistent with the logic of relativity, the capuchins working for cucumbers often refused to participate further when another monkey was paid in grapes for the same activity. De Waal shows a video from this experiment in his TED Talk *Moral Behavior in Animals*, and I highly recommend it, if only to see an irate monkey chuck his cucumber slice at the experimenter after the monkey next door receives a grape. At the risk of anthropomorphizing, I've never seen a fellow primate more outraged at unfair pay than that little capuchin.

Experiments such as this one provide compelling evidence that once your survival needs are met, everything else is relative. As we discuss in chapter 3, life was a zero-sum game in the world before cities. My fortune typically came about only as a result of your misfortune. The sad truth is that the logic of sexual selection ensures that life remains a zero-sum game even when it doesn't need to be. If my friend gets a big raise or wins the lottery, I am indeed impoverished by his good fortune, because now it'll be harder for me to get a mate. Prior to his achievement or good fortune, the woman I had my eye on might have chosen me. Now there's a good chance she won't, as he's more attractive than he used to be.

Sexual selection etches this logic deep into our psyche, making it incredibly difficult for us to rise above it and not feel envy at the success of others—particularly the success of close others. When Sam Smith or Eminem wins a Grammy, I don't get envious because I don't know them, I don't run in their crowd, and I (wisely) don't try to compete with them. The same holds when Leonardo DiCaprio wins an Oscar; his girlfriend wasn't going to go out with me anyway. Yet it hits close to home when my good friends outperform me, particularly in a domain I value.

My favorite demonstration of this effect is a wonderful series of experiments by Abraham Tesser of the University of Georgia and his colleagues. In a typical experiment, Tesser brought male college students into the lab to play a word game. The game was a dictionary-style task in which people passed clues to help their partner guess a target word. Tesser arranged for one participant to go first and rigged the game so that the first player always performed poorly. The key issue concerned the quality of clues that this ostensibly poor-performing participant then went on to give his partner. Some of the clues he could give his partner were designed to be really helpful, and some were designed to be really unhelpful.

To make this experiment concrete, imagine you were participating and your task was to help your partner guess the word *insightful.* If you wanted to help him out, you might give him the clue *astute* or *shrewd.* But if you wanted him to fail, you might give him the clue *perspicacious.* (Who knows what that means?) Later, if he were to complain, you could point out that if he had a half-decent vocabulary, he'd have nailed it.

Tesser found that people were more likely to give helpful clues to their partner if he was a friend rather than a stranger, but this result held only when the task was described as unimportant. When the task was described as an important indicator of verbal skills, Tesser found that after failing at the task themselves, people were more likely to give helpful clues to strangers than friends. These data suggest that good performance on the part of a friend was more threatening than good performance on the part of a stranger. It's an ugly truth about people, but these results show that we undermine our friends when there is a risk they'll outperform us in important domains. This is what sexual selection has wrought.

Part II

Leveraging the Past to Understand the Present

5

Homo Socialis

A few decades ago I was teaching with Semester at Sea, a study-abroad program that travels the world by ocean liner. One moonless night on the Indian Ocean, the captain switched off the exterior lights so we could more easily see the stars. I was curious to see the Milky Way without any light pollution, so I meandered up to the top deck before going to bed. The night sky was crammed with stars, and I had spent about five minutes admiring it when an extraordinarily bright meteor blazed into view and then burned out right before my eyes.

It was one of the most amazing sights of my entire life, and I found myself surprised by my own response. Rather than savoring the moment or reflecting on my good fortune, I immediately looked around to see if anyone else was on deck with me. All I wanted was to turn to someone, anyone at all, and say, "Wow, did you see that?" and hear even a complete stranger respond, "Yes. That was amazing!" As a child, I had always envied Tigger in *Winnie the Pooh* for being the *only one*, but this experience taught me that I misunderstood my own preferences. Rather than feeling more special because I was quite possibly the only human on earth to witness this spectacle, I found that the meteor felt less real and the experience of seeing it less meaningful because I couldn't share it.

Of all the preferences that evolution gave us, I suspect the desire to share the contents of our minds played the single most important role in elevating us to the top of the food chain.* We are the fiercest predator on the planet by virtue of the power of our minds, but even human minds aren't that special on their own. If you drop one of us naked and alone into the wilderness, you've just fed the creatures of the local forest. But if you drop one hundred of us naked into the wilderness, you've introduced a new top predator to this unfortunate stretch of woods.

In chapters 1 and 2, I discuss the central role of social functioning in our evolutionary success story. Nothing is more important to us than our social connections because nothing was more critical for our ancestors' survival and reproduction. As a result, we've evolved many ways to stay connected to our groups; chief among them is to know what others are thinking. Knowing others' thoughts helps us fit in and predict what our group members will do next. We also want our group to know our thoughts and feelings, as planting our beliefs in the minds of others provides the best opportunity for nudging the group in our preferred direction. Acceptance of our thoughts and feelings by others also validates our place in the group and gives us a sense of security about our future. As luck would have it, these two inherently self-serving goals are also a recipe for successful cooperation; when we know the contents of one another's minds, we are much more capable of social coordination and division of labor.

For these reasons, evolution has given us a perpetual desire to share the contents of our minds, even when there is nothing to gain at the moment by doing so. This desire to share our experiences, which I felt so acutely on the deck of the ship that night, emerges early in life. Toddlers endlessly narrate the world, pointing out peo-

* The desire to share the contents of our minds is one of two key traits that Thomas Suddendorf identifies as uniquely human in *The Gap*. More on this in chapter 6.

ple and objects simply to establish joint attention. No other animal does this at any stage of development.

The desire to share our understanding and experience goes beyond mere knowledge, as we also want to share our emotional reactions with others. For our group to deal effectively with a threat or opportunity, we must all perceive it the same way, and thus we have evolved to seek emotional consensus. There are few experiences in life more frustrating than sharing an emotional story with someone who reacts with indifference or an emotion opposite to our own. If I'm outraged by my colleague's rude behavior, I'm even more outraged when my wife thinks it's no big deal or, worse yet, funny or justified.

This need to share emotional experience sits at the root of nearly all exaggeration. If I'm worried that you won't be sufficiently impressed by the fish I caught, then the fish grows in the telling. If I'm worried that you won't be upset by my colleague's rude behavior, then my colleague gets ruder when I relay the story. This need also sits at the heart of urban legends, which are the viral equivalent of gross exaggerations.

Chip Heath of Stanford University and his colleagues provide a nice example of this effect in their research on urban legends. They found that the more disgusting and outrageous the legends become, the more likely people are to say they would pass them on to others. For example, they presented people with different versions of an urban legend in which a family returns from vacation to find photos in their camera that the bellboy took of himself fouling their toothbrushes. In one version, the photos show him cleaning his nails with their toothbrushes, in another he has their toothbrushes stuck in his armpits, and in the coup de grace, they see the bellboy with the family toothbrushes "up his bootie." It doesn't take much thought to decide which story you'd tell your friends; the family toothbrushes in the bellboy's bootie are a clear winner if your goal is to ensure your audience shares your emotional reaction.

When we exaggerate or pass on urban legends, we introduce distortion into our listeners' understanding of reality. This can be seen as a costly side effect of the need to share emotions, but it shouldn't detract from the fact that sharing emotions lies at the root of almost all successful social interaction. Think back to the last conversation you had with a friend and ask yourself how much new information was transmitted versus how much emotion was transmitted. Meaningful conversations are often very light on information, but they are almost never light on emotion. And conversations that are heavy on content, such as discussions of current events, are typically laden with emotion as well. The rare conversations that involve no emotion at all are typically with strangers and are perceived as trivial or boring.

Social Intelligence

The most important challenge people face in life is understanding and managing others. If I understand other people's goals, I can position myself to benefit from their likely actions. Better yet, if I can manage other people's agendas such that I plant my goals in their minds, I will almost assuredly be a success in life. On the other hand, if I can't understand others, I'll be buffeted around by their seemingly random plans. And if I can understand others but not manage them, I'll see the bad news coming but will have limited capacity to improve my situation. Management skills are less important if I'm Genghis Khan and can push my agenda on others by brute force, but for most of us, persuasion is the key to success.

What does the science tell us about how to be a social success? Unfortunately, not much. The answer has proven to be incredibly elusive due to the challenges of measuring social intelligence, which have remained largely unchanged since the first comprehensive test was created in the mid-1920s. Consider an item on that original test that asked people what they would say to an acquaintance whose relative had just died. Participants were presented with the choice of speaking well of the departed relative or talking about current events of general interest. A moment's reflection makes it clear that either or neither of these choices might prove to be correct.*

Indeed, the "correct" response to this question depends on how well you know the acquaintance and the deceased, their relationship, and countless other factors. As a consequence, one person's gaffe could easily be another person's comforting or uplifting words. Although it's generally inadvisable to make jokes about the deceased at the funeral of a beloved relative, I suspect that many people have in fact comforted their friends and family with such jokes. And I'm equally confident that many other people have offended their friends and family with the exact same jokes. The same statement can be either comforting or upsetting depending on who makes it, who is listening, exactly when it is made, the tone of voice with which it is made, and so on.

The more you think about this problem, the more you realize that there is rarely a single correct social behavior in any given situation. The context-dependent nature of emotional responses, and hence social appropriateness, continues to stymie our efforts to measure social abilities. For example, an item from one of the most widely used current measures of social/emotional intelligence describes a person who works harder than his colleagues, gets better results, but is not good at office politics and loses out on a merit award to an undeserving colleague. Test takers are asked how much

* For what it's worth, the authors of the original test declared that talking about current events is the correct answer. So, I googled "What are you supposed to say at a funeral?" and three of the first four answers involved speaking well of the departed relative (the fourth was about putting the family in your prayers). Perhaps norms have changed over the last ninety years, but I suspect there never was and never will be a universally correct way to handle most social situations.

it would help the person who lost out on the award to feel better if he made a list of the positive and negative features of his undeserving colleague versus told others what a poor job his undeserving colleague had done and gathered evidence to prove his point. In truth, the answer to such a question is unknowable. Some people would benefit from one strategy, some from another, and the benefits they'd gain would assuredly depend on a wide variety of other factors and constraints. The latter strategy could clearly be a disaster, but it may conceivably be a winner as well.

The developers of the test provide "correct" answers by assessing the consensus of what most people would do, and by surveying experts in emotion research. Both these approaches are deeply inadequate, however, as they rely on the flawed assumption that there is a single correct answer waiting to be found. Even if such an answer existed, this approach also assumes that what most people would do is the best strategy, thereby limiting the upper end of social intelligence to the average response (or, perhaps even worse, the intuitions and inclinations of academic psychologists). I suspect that highly socially skilled individuals achieve their goals in part by doing things differently. When most people respond in a certain way, that response loses some of its power through repetition and predictability. Socially skilled people recognize this problem, and by finding a way to be just a little different from everyone else, they communicate more effectively.

These challenges struck me as insurmountable, so they led me to take a different approach to the problem. Rather than trying to find a way around the context-dependent nature of social intelligence, my colleagues and I decided to harness this defining feature of social functioning. There are clearly many qualities that enable people to be socially successful, but the fact that what works in one situation often does not work in another suggests that *behavioral flexibility* may be the single most important attribute for skilled social functioning. A number of factors enable behavioral flexibility (more on this later), but one of the most important is our capacity for self-control.

The Evolution of Self-Control

A running joke among academic psychologists is that we entered this field to try to understand our own failings—that is, we conduct "me-search" rather than research. I'm as guilty of this charge as the professor in the office next to me. My interest in social intelligence began not with a burning intellectual question, but with an embarrassing faux pas when I was in high school and blurted out the first thought that passed through my head.

Speaking before thinking gets me into trouble in a variety of circumstances, but food markets are my Achilles' heel, particularly when meat looks like it did on the hoof. In combination with a fair bit of squeamishness, my carnivore's guilt over the fact that my dinner once roamed this planet leads me to prefer foods that don't resemble the animal from which they came. I get the willies when tapas bars hang pig legs from the rafters, when Chinese restaurants hang ducks in rows on hooks, or when a butcher's shop displays an entire carcass. I don't even like fish looking back at me. More than once I've made a disgusted expression in front of a shopkeeper before my (admittedly weak) internal filter had a chance to stop me. My friends and family are unimpressed with my behavior, so I set out to prove that it's not my fault, that I'm the victim of faulty brain structures and should be pitied rather than censured.

One way to envision self-control is to imagine that your brain is like a chariot. The horses are your impulses, and they reside primarily in a small set of regions that sit under your cortex, near the base of your brain, such as the nucleus accumbens and the amygdala. The horses pull you toward gratification of your desires: food, sex, aggression, whatever it may be. Some people have wild stallions pulling their chariot, and they struggle to resist the temptation to eat too much, drink too much, have affairs, or punch the annoying guy in the nose. Other people are pulled along by petting farm ponies, and for them, managing their impulses is comparatively easy.

The chariot driver, who sits in a piece of the frontal lobes called the lateral prefrontal cortex (LPFC), is the one who resists temptation by reining in or redirecting the horses when the time, location, or goal itself is inappropriate. The driver has a copilot, who sits above the horses in or about the anterior cingulate cortex (ACC), and whose job is to alert the driver whenever the horses appear to be heading in the wrong direction. If you have an inattentive copilot or a weak driver, the horses will pull you pretty much wherever they like, and people will tell you that you're a wild one or an impulsive jerk (depending on whether they sympathize with your actions).

In my own case, I think my ACC copilot is on permanent vacation. Perhaps my ACC is too small or is chronically starved of oxygen, or maybe it's just too quiet to be easily heard by my driver. Once my LPFC is pulling the reins, I've got perfectly good self-control, but I often don't notice that I need to control myself until it's too late. I blame my ACC, but of course others blame me. So my first experiment on social intelligence was an exculpatory effort designed to test my hypothesis that blurting out what you're really thinking is a sign of faulty frontal lobes rather than a moral failing.

In this experiment, we decided to reenact my frequent faux pas with unusual foods, which meant that we had to come up with an excuse to present people in the lab with a food item that looked as it did when it originally walked this planet.* After putting our heads together, my doctoral student Karen Gonsalkorale and I decided to use her Chinese ethnicity and cooking skills to our advantage. We brought Caucasian participants into the lab, and Karen explained—that is, she lied—that she was testing the effects of different food chemicals on memory.

* And we're back to the me-search component of my work.

After looking up the participant's name and pretending to consult her clipboard, Karen then told each person, "You're in luck! You get to eat my favorite food, which is widely regarded as the national dish of China!" (Our deception regarding food chemicals and memory allowed us to serve the participants an unusual food in the lab that would otherwise have seemed odd.) Karen's claims about the personal and cultural significance of the dish were intended to communicate one simple message: whatever you're served next, you should at least pretend to like it.

On hidden camera and in close proximity to the participants' faces, she then opened a Tupperware bowl containing intact chicken feet, claws and all, cooked in a light brown sauce. Not everyone responded politely when given the feet to eat. My favorite participant blurted out, "That is bloody revolting!" This statement was followed by an awkward silence, which he broke with a few apologetic mumbles as he sheepishly picked up a foot and tried to muster the courage to nibble on one of the toes. In contrast, some people never lost their composure. They didn't necessarily eat the feet—many suddenly remembered they were vegetarian or suggested that the feet might not be kosher—but they were polite even if they declined to eat them.

In the next phase of the experiment, we used the Stroop test to see if we could differentiate those who had handled this delicate situation politely from those who had not. The Stroop test takes advantage of the fact that reading is automatic, and thus we cannot help but read a word once we see it. For example, try to look at the word below without reading it:

Hello

I'd hazard a guess that you failed—if you saw it, you read it. In the Stroop test, people are asked to name, as quickly as they can, the colors of the letters that make up various words. The tricky bit is that the words themselves are names of colors. For example, people might see the word *red* written in green letters, which requires them to inhibit their automatic tendency to report the word they read and instead report the color of the letters.

When people take the Stroop test in an fMRI magnet,* we see that the ACC copilot becomes active whenever the color word doesn't match the color of the letters and there is competition in people's minds between the correct and incorrect response. This ACC activation is even stronger when people make a mistake. At that point, the ACC alerts the chariot driver, who reins in the tendency to read the word rather than name the ink color.

Because the Stroop test taps processing in the ACC, we expected that it would predict how people reacted to the chicken feet. People who have a responsive ACC should be good at inhibiting their initial reaction (*yuck!*) and replacing it with something more socially appropriate (*interesting!*). Consistent with our expectations, those who did well on the Stroop test were less likely to lose their composure than those who did poorly on the Stroop test. These findings showed us that people who had better self-control were able to respond in a more flexible and socially skilled manner. They were able to inhibit responses that they would typically have made but were inappropriate under the current set of social rules.

Subsequent work has shown that the blurting we found in the lab is only the tip of the iceberg. Two years later, Christopher Patrick of the University of Minnesota and his colleagues showed that a quiet ACC not only is problematic when it comes to *saying* the wrong thing; it also stands idly by when people *do* the wrong thing. In their study, they selected people who either reported having engaged in previous antisocial behavior or not and gave them a task very similar to the Stroop test while they wore a swim cap covered with electrodes. Because neurons emit tiny amounts of electricity,

* An instrument that allows researchers to monitor metabolic activity in different brain regions.

these electrodes enabled the researchers to see how their participants' ACC responded when they made a mistake and pushed the wrong computer key.

Despite the rather bland nature of this laboratory task, Patrick and his colleagues could tell whether someone had engaged in antisocial behavior by how their ACC responded when they made a trivial mistake. When they made an error on this task, the more law-abiding folks showed an ACC response that was about 30 percent larger than that of the antisocial folks. Remember, the ACC copilot's job is to notice the potential for conflict and alert the driver when an error is coming. A poor copilot responds weakly to the possibility of error, much as we see in this experiment. Presumably when the antisocial people were about to make a more important decision (such as whether to punch the guy who's annoying them or throw a rock through someone's window), their ACC was similarly quiet and failed to alert them to the conflict between what they were about to do and what they should be doing. These data reveal how an unresponsive ACC copilot can lead to poor social functioning.

I was delighted that these experiments vindicated me, showing that my failings revealed faulty brain structures rather than a faulty character. In retrospect, though, it's clear that my self-serving agenda blinded me to the more important message this research was telling us. These studies show that self-control plays an important role in social functioning, but it didn't occur to me for quite some time that social demands are probably what led to the evolution of self-control in the first place.

You don't need an ACC copilot to tell you not to grab a salmon from a bear's mouth; that would be obvious even to the most distracted of chariot drivers. Rather, you need your ACC to ring the alarm when you're reaching for the last piece of cake, flirting with the big guy's girlfriend, or starting to tell your boss what you really think of him. Social interactions are loaded with conflicting motives, which is precisely when we need the services of an attentive copilot. My goals are sometimes consistent with yours (e.g., when we want to see the same movie) and sometimes at cross purposes (e.g., when we want to see the same girl), and that's when an attentive copilot plays a particularly valuable role.

It was not just my egocentrism that blinded me to what these studies showed. Most psychologists have taken it for granted that we evolved the capacity for self-control in order to pursue long-term goals. To be a successful farmer we must plant the seed rather than eat it; to have a happy retirement we must save our money rather than spend it; to maintain a healthy body weight we must decline the second piece of chocolate cake rather than eat it. But our world looks nothing like the world of our hunter-gatherer ancestors, who didn't plant seeds or save money and who never worried about eating or drinking too much. Our ancestors were focused on today, with occasional thoughts of what they'd like to do tomorrow, so their lives were not the perpetual exercise in delayed gratification that ours have become. They almost assuredly did not rely on self-control to secure a better tomorrow. But they did have to control themselves to get along with their neighbors, manage their rivals, and achieve their social goals.

Our ancestors also had to control themselves whenever they engaged in cooperative efforts, particularly in the face of threat. Imagine what it was like for our distant ancestors who had migrated to the savannah and were a toothsome snack for hyenas or lions. When they noticed one of those beasts stalking them, it would have taken a monumental act of self-control to band together and throw stones at it rather than run away. A cooperative strategy was clearly the most effective way to achieve the survival goal that everyone shared, but fear would have made it awfully tempting to leave the task of protecting the group to others. Those who were unwilling to share the burden and who ran away at the first sign of danger would soon have found themselves unwelcome in the group, facing dire circumstances and poor reproductive prospects. Evolution thus favored the development of self-control in our ancestors to achieve a variety of social goals.

The joint actions of the ACC and LPFC enable self-control, but there is more to self-control than restraining ourselves in the face of temptation. Our big brains also allow us to reframe the physical world in abstract terms, helping us to see it in terms of problems to be solved rather than potentially overwhelming temptations. To explain what I mean, let's consider a wonderful experiment with chimpanzees conducted by Sarah Boysen and Gary Berntson at Ohio State University.

First Boysen taught the chimps the numerals for one through nine. She then taught them to play a game that involved choosing the number of treats they would like to receive. In the game, two chimps are seated across from each other as in Figure 5.1a, and one chimp—we'll call this one the chooser—has the opportunity to decide how many treats each will receive. The chooser is shown numbers on two separate cards, and its task is to point to one of the two numbers.

The trick to this game is that the other chimp receives the number of treats shown on the card the chooser chimp points to, and the chooser receives the number of treats indicated on the *other* card. Chimps are not good at sharing, so their goal is always to get the larger pile for themselves and have the smaller pile go to the other chimp. As a consequence, they soon learned to point to the smaller number in order to score the larger pile of treats (see Figure 5.1a).

The key to the study was what happened when Boysen presented the chimps with actual piles of treats rather than cards with numbers. Actual treats should have made the task easier, and the chimps more successful at getting the larger pile, because the chimps don't need to remember what the numbers represent to

Figure 5.1a. The chimp plays the game well. By pointing at the smaller number, the chimp gets the larger one. (Sarah T. Boysen)

win the game. Yet the exact opposite occurred; despite knowing the rules of the game, the chimps repeatedly pointed to the larger pile of tasty treats in front of them. The chimps appeared to realize their mistake immediately, often seeming very frustrated with themselves. Remarkably, they went on to make the same mistake on the very next round (and the next and the next, for literally hundreds of rounds).

Why did these clever animals repeatedly make this simple error? Most likely they could not escape the lure of the actual treats. When Boysen helped them translate treats into an abstract problem by showing them numerals, the chimps were able to take a mental step back and consider the problem objectively. But their limited symbolic abilities weren't strong enough to transform the physical treats in front of them into an abstraction, and as a result, the temptation was too great. The treats were just too enticing for the chimps' frontal lobes to put the brakes on and stop them from pointing to the larger pile.

Figure 5.1b. The chimp plays the game less well. By pointing at the larger pile of treats, the chimp gets the smaller one. (Sarah T. Boysen)

Because chimp frontal lobes are smaller than our own, their control functions and ability to think in abstract terms are not as strong as ours. Humans can easily convert piles of candies into an abstract problem, allowing us to think of the treats as numbers rather than objects. Once we translate our problems into abstractions, our temptations don't swamp our control functions. In contrast, the chimps could achieve this conversion only when Boysen did it for them, by presenting them with numbers rather than actual candies. Lest we feel too superior to these beasts, it's worth remembering that we are a lot like chimps when we're children, with weak frontal lobes and limited abilities of abstraction. As a consequence, we can see this same process in action among (young) humans, most notably in Walter Mischel's famous marshmallow studies.

Mischel brought young children into the laboratory and sat them down in a room with a single marshmallow on a tray in front of them. He told the children that they could eat the marshmallow now, if they'd like, but if they waited until he returned, he would give them *two* marshmallows. He also told them that if they decided they couldn't wait, they were to ring a bell, and then they could eat the marshmallow. It won't surprise you to learn that the children almost universally wanted two marshmallows and vowed to wait until Mischel returned. He then left the room and, unless summoned by the bell, did not return for fifteen minutes.

Mischel wanted to know how long the kids could wait and what would predict their self-control. Several interesting results emerged from this study. First and foremost, there were huge individual differences in how long the kids could wait—many held out for the entire fifteen minutes, but some barely made it to the ten-second mark. If you watch the tapes from the original experiments, one factor stands out in predicting who waits and who doesn't. The kids who are able to wait the longest are the ones who turn their attention away from the marshmallow. They sing songs to themselves, turn their backs on the tray, play games, and even fall asleep. But the kids who stare at the marshmallow or, worse yet, hold it in their hot little hands have no hope. Down the hatch it goes.

Resisting marshmallows is particularly difficult for preschoolers, as their frontal lobes are underdeveloped, so their self-control skills are limited. But some kids figured out how to work around their own weaknesses, in much the same manner that Boysen was able to help her chimps by translating the physical candies into numerical abstractions. When Mischel tracked down these same children a dozen years later, he found that the kids who waited longer to eat the marshmallows had better SAT scores than the kids who folded quickly. Their skill in translating the temptation into a problem they could solve helped them resist temptation when they were small children and continued to help them engage in self-control throughout their life, presumably enabling them to study more and party less.

Beyond Self-Control: Social Benefits of a Big Brain

About fifteen years ago, I learned of the social brain hypothesis, an idea that had been knocking around in biology and anthropology since the 1960s but was largely unexplored in psychology. As I discuss in chapters 1 and 2, this hypothesis posits that primates evolved large brains to manage the social challenges inherent in dealing with other members of their interdependent groups. It eventually occurred to me that if our brains grew so large to solve social rather than physical problems, then many of the abilities that we regard as purely cognitive might play an important social role.

For example, perhaps we evolved the capability to conjure up alternative approaches to problems (known as divergent thinking) not to find a way across a raging river or to escape hungry hyenas, but to facilitate flexibility in social situations. Divergent thinking would have allowed us to deal more effectively with friends and enemies when our initial approach failed. This idea resonated with me because it fit my personal experience growing up. As a particularly small kid with a particularly big mouth, I relied heavily on my divergent thinking skills to extricate myself from dicey situations on the playground.*

To test the idea that divergent thinking enhances social success, my PhD student Isaac Baker brought groups of friends into the lab and ran them through a series of tasks. He tested their IQ, measured their personality, and asked them how many different uses they could think of for a brick, a plate, and other common objects.

* Although I remember it fondly, any modern visitors to my Alaskan schoolyard of the 1960s and '70s would think they had landed at Our Lord of the Flies Elementary. The common assumption among our teachers was that the subzero temperatures would clot our bloody noses by the time recess ended, so they let us resolve our differences as we saw fit. Those school days taught me a great deal about the life of a hunter-gatherer; we, too, had to get by on our wits, with little hope of intervention by an impartial authority.

These latter questions tap divergent thinking, as some people will give answers that are pretty similar to one another (e.g., use a brick to hold open a door, hold open a window, prop up a shelf), and others will show a wider variety in their approach (use a brick to weight the corner of a picnic blanket, hammer a nail, throw at an annoying person).

He then asked everyone to privately report on the social skills of the other members of their friendship group. Isaac found that people who came up with more divergent uses for a brick were also more persuasive, humorous, and charismatic. This relationship held whether they had a high IQ or not, so it wasn't just the case that divergent thinking was a reflection of being a smart person. Rather, divergent thinking is an important skill in its own right, enabling people to be more persuasive, amusing, and charismatic.

Mental speed, or the ability to retrieve information and solve problems rapidly, is another cognitive capacity that enables a flexible response to the world. Because social interactions often move quickly, they provide very little time to think. If you make a joke at my expense and I immediately come back with a witty response, I've held my own in our little banter. But if it takes me too long to come up with a response, then the conversation is likely to have moved on. By then, even if I've thought of a brilliant retort, I'll look like a fool if I try to respond to your earlier point. The faster I can think, the broader the array of options I can consider before I have to respond.

To examine whether mental speed predicts social functioning, we conducted a study in which we asked people in friendship groups to answer simple common-knowledge questions (e.g., "Name a precious gem") as quickly as they could. We then asked their friends how charismatic they were, and found that people who answered the common-knowledge questions more quickly were rated by their friends as being more charismatic. And as with divergent thinking, these effects of mental speed were independent of intelligence.

Research over the last century has taught us that IQ is our intellectual horsepower, and that social intelligence is just a subset or offshoot of that larger class of mental abilities. In contrast, the results of these initial studies suggest that perhaps we have it backward; maybe our social intelligence is our true intellectual horsepower, and our ability to solve complex problems (i.e., abstract intelligence/IQ) is just a fortuitous offshoot of our evolved social capacities. If we take the social brain hypothesis seriously, it suggests that IQ is a by-product of social intelligence rather than the other way around. And if social intelligence really represents our broader mental abilities, then it makes perfect sense that IQ is often a poor predictor of career success. When we measure IQ, we're looking at only a small slice of the cognitive pie, while our social intelligence might tell us much more about our capacity to navigate the world.

At this point I can easily imagine a counterargument along the lines of "Hold on. Many of the smartest people I know are social misfits, while some of my socially skilled friends can't add up their grocery bills. If IQ is an offshoot of social intelligence, shouldn't the two be more highly related?" Such discrepancies are all too common, and they clarify that there is no one-to-one correspondence between overall cognitive abilities and social abilities. This is precisely why we have been exploring particular cognitive capacities (such as self-control, divergent thinking, and mental speed) that enable flexibility and thus should make people more socially skilled. In the same manner that some people can memorize the Constitution but can't find their way home from the supermarket, some people are great at math but don't have the particular cognitive skills that help them understand and manage others.

Finally, it is important to note that social skills depend on more than just cognitive abilities; they also depend on our attitudes. Perhaps surprisingly, one of the most important social attitudes is the one we hold toward ourselves.

The Social Benefits of Overconfidence

The *von* part of my surname reveals my German heritage—if you went back far enough, you'd find that my father's family were landowners under the kings of Prussia. That means we have a family crest (which happens to resemble an advertisement for St. Pauli Girl beer) and a family motto, *Mehr sein als scheinen*. Translated literally, it means "More to be than to seem," which I understand as "Be more than you seem to be." A Google search suggests that our family motto is shared by many of our erstwhile neighbors, as modesty is a Prussian virtue. "Be more than you seem to be" is an excellent approach to life if you're a ninja or a card shark, but for almost everyone else our motto has it exactly backward. We get more in life if we seem to be more than we are. Indeed, I strongly suspect that I regard myself as "Bill plus 20 percent." Let me explain.

In one of my favorite studies on overconfidence, Nicholas Epley of the University of Chicago and Erin Whitchurch brought people into the lab to have their pictures taken. Epley and Whitchurch then morphed these photographs to varying degrees with attractive or unattractive photos of individuals of the same sex as the participant. A few weeks later, Epley and Whitchurch asked their participants to return to the lab and presented them with either morphed or unaltered photos of them under different circumstances. In one experiment, participants were asked to find their true photo in a jumbled array containing their actual photo and the various morphed photos of them. In that experiment, participants were most likely to guess that their true photo was the one that was morphed by 10 to 20 percent with the more attractive image.

In a second experiment, participants were presented with an array of photos of other individuals, among which was a single photo of them that was either unaltered or morphed 20 percent with an attractive or unattractive image. In this experiment, Epley and Whitchurch found that people were able to locate photographs of themselves most rapidly if those photos were morphed with an attractive photo, at an intermediate speed if they were unaltered, and most slowly if they were morphed with an unattractive photo. These findings suggest that the enhanced photo most closely matches how people see themselves in their mind's eye, suggesting that we deceive ourselves about our own attractiveness.

If you think you would be immune to this self-enhancement effect, consider for a moment the last time you saw a candid photo of yourself that you liked. If you're similar to most people, you probably think that candid shots of you are typically poorly taken. In my own case, I'm confident that my friends aren't just bad photographers but downright sadistic. The sad truth is that our friends are not bad photographers; we're just not as good looking as we think we are. And that's why you don't like candid pictures of yourself: because they capture what you actually look like, not what you think you look like. You prefer the picture of yourself that caught you at just the right angle, on just the right day, and those are the ones you put up on Facebook, Tinder, or in the company directory. Because you then see this photo more often than you see the (deleted) ones you disliked, you also come to believe that this unrealistically good picture is an accurate representation. No wonder you end up thinking of yourself as "you plus 20 percent."

The Epley and Whitchurch study is a great example of self-deception in action, but it doesn't tell us *why* people deceive themselves, a question that has been debated at least since Socrates and Plato. Socrates was a big fan of the Delphic admonition to "know thyself," and he delighted in demonstrating to the Athenian elite that they didn't know as much as they thought they did. His fellow Athenians didn't appreciate his propensity to highlight their inadequacies, and they famously sentenced him to death on trumped-up charges. As a psychologist, I, too, think that self-knowledge is often highly overrated. All it takes is a glance at a photograph of my adolescent self to realize the enormous damage that self-knowledge can cause. If I had any idea what a twerp I looked like in ninth grade, I would have struggled even to go to school, let alone flirt with the girl in the seat next to me.

Freud also saw the value of self-deception, believing that it is intended to protect us from a world that is often too unpleasant to bear. Perhaps there is some truth to this, as it might be hard to face another day if we knew what our friends and neighbors really thought of us. But as I discuss in chapter 9, evolution didn't design us to be happy; it designed us to be successful. And it's easy to imagine how a misguided sense of self would make us a lot less successful. Bill plus 20 percent is going to get in fights he'll lose, and is going to ask out women who are disinterested (and, if I remember right, laugh in his face). So how do the gains offset the costs?

Although psychologists ignored his insight for nearly forty years, Robert Trivers proposed a simple and brilliant answer to this question when he was a young professor at Harvard in the mid-1970s. He suggested that we deceive ourselves in order to deceive others more effectively. If my ninth-grade self (incorrectly) believes that I'm not a twerp, when I ask out the pretty girl in my biology class she'll be faced with a conundrum. On the one hand, I certainly don't look like much. On the other hand, I seem to think otherwise, so perhaps there is more to me than meets the eye.

Overconfidence can thus be beneficial if I can plant my inflated self-views in other people's minds. And if I'm particularly good at planting my inflated self-views in the minds of others, I might not need to pay the costs I've just highlighted (getting my nose smashed or looking like a fool). The guy who can beat me up might wonder if I'm tougher than I look and think that perhaps he should let the matter drop, and the woman who can do a lot better might be tricked into thinking that "a lot better" is me. After all, I've known myself much longer than they have, so it would be foolish to ignore my opinion of myself.

There are only a few tests of this idea, but so far, the results are consistent with Trivers' hypothesis. For example, Cameron Anderson at Berkeley and his colleagues put students in small working groups and found that they couldn't differentiate between their peers who were knowledgeable and those who were overconfident. As a consequence, they tended to defer to overconfident people when they shouldn't. Richard Ronay at the University of Amsterdam and his colleagues found similar effects among human resource consultants who were tasked with determining which applicants should be promoted to managerial positions. The HR consultants tended to recommend the overconfident applicants instead of the applicants who were better calibrated in their self-knowledge. Just like Cameron Anderson's students, even trained HR consultants couldn't differentiate between people who really knew what they were talking about and people who were just blowing hot air.

My PhD student Sean Murphy went on to show that these effects are not just evidence that people can be tricked in the short term by those they don't know well. Sean found that high school boys who were overconfident about their sporting ability actually became more popular from one school year to the next. These data suggest that overconfidence not only is effective when people don't know you well, but also has a positive impact in long-term social networks. Finally, and perhaps relatedly, in another set of experiments, Sean found that one of the reasons that overconfident people are successful is that they are more intimidating in competition, so people don't like to go head-to-head with them.

These studies suggest that there are notable interpersonal benefits to overconfidence, and that Freud was wrong when he characterized self-deception as a defense mechanism. Rather, Trivers hit the nail on the head when he suggested that self-deception is more aptly described as a social weapon. Bill plus 20 percent isn't trying to protect his psyche from an admittedly inhospitable world; he's trying to get people to like him and avoid conflict with him. To put it more generally, my tendency toward self-inflation evolved to help me achieve social outcomes that I couldn't get if I were honest about who I truly am.

Self-Deception Isn't Just for the Overconfident

Trivers' insight that self-deception is a weapon of social influence and not a strategy for feeling better helps us understand overconfidence, but self-deception goes well beyond that. Our self-confidence is not the only factor that is readily perceived by others, as nearly all our emotions have social consequences. Let's consider one of our most important emotions: happiness. I spend plenty of time in chapters 9 and 10 talking about what makes us happy, so for now, the important point concerns the social consequences of happiness. A moment's reflection reveals that happiness has substantial social impact, largely because we enjoy spending time with happy people. As the aphorism goes, "Laugh, and the world laughs with you; weep, and you weep alone."

The social consequences of happiness are reason enough to put on a happy face, and people exaggerate their happiness in numerous social interactions. If I run into you on the street and ask, "How are you doing?" in truth, I don't want to hear about your bunions or hemorrhoids, but rather I'm just hoping you'll say, "Good! How are you?" Even when our bunions or hemorrhoids are bothering us, in brief social interactions we almost always tell other people that things are fine. But the effect goes deeper than just telling people one thing while believing another. If Trivers is right, we actually try to convince ourselves of the truth of these claims in order to persuade others. We can see this possibility in Epley and Whitchurch's study with the morphed faces, Anderson's study with the overconfident students, and numerous other studies in which people overestimate themselves and potentially lead others to do the same. So how might such an effect emerge in the case of happiness?

My favorite study on overhappiness is by Sean Wojcik of UC Irvine and his colleagues. The background to their study is an effect that is well known among social scientists: people on the conservative side of the political spectrum in the United States tend to be happier than people on the liberal side. A number of hypotheses for this effect have been proposed, but Wojcik and his colleagues had a different idea. When they dug into the literature, they realized that the data showed that conservatives *claim* to be happier than liberals, but no one knew whether they *acted* any happier. So, Wojcik took a deep dive into the kinds of big data that are now publicly available via Twitter, LinkedIn, and the Congressional Record. He pulled out millions of words, thousands of tweets, and hundreds of photographs from members of Congress and other people known to be on the political left or right, to see if there really are differences in the positivity of their language and the size of their smile.

The first question you might ask is why would conservatives claim to be happier than liberals if they really aren't? If you reflect on the ideology of modern American conservatism versus liberalism, one of the clear differences between the two groups lies in their beliefs regarding the fairness of the playing field. Conservatives more strongly endorse the idea that the world is a meritocracy than liberals do, as liberals tend to see a variety of structural barriers to achievement that conservatives regard as relatively unimportant. For example, liberals believe that one's race, sex, or sexual orientation can lead to unfair treatment and loss of opportunities, whereas conservatives tend to believe that the effects of race, sex, or sexual orientation are largely overblown.

If you follow these beliefs to their logical conclusion, you'll see that conservatives should believe (consciously or otherwise) that people are more responsible for their own happiness than liberals do. If I'm an unhappy conservative, and if the world is a meritocracy, then I must have failed to achieve my goals, or otherwise I would be happy. In contrast, if I'm an unhappy liberal, then it's very possible that my race, sex, social class, or something else about me that I cannot control has held me back, so my unhappiness is not necessarily a sign of my own failings. For this reason, it's more important for conservatives to claim happiness than it is for liberals, as lack of happiness suggests failure on the part of conservatives but not necessarily on the part of liberals.*

Consistent with this logic, when Wojcik examined the things that liberals and conservatives say or tweet, and the way they look in their photos, no evidence emerged that conservatives are happier than liberals. In fact, Wojcik found just the opposite. Liberals used more positive words and showed larger smiles in their photographs. Smiles not only differ in size, but genuine smiles can also be differentiated from posed smiles by the crinkles that appear around your eyes. Genuine smiles almost always lead to eye crinkling, but posed smiles often don't. When Wojcik and colleagues coded the photographs for the presence versus absence of eye crinkling, they found the liberals were more likely than the conservatives to show this evidence of genuine happiness.†

* At least that is what it suggests to their fellow conservatives and liberals, who tend to be more important peers than members of the opposing party.

† I have no idea why liberals would be happier than conservatives, but there are lots of possibilities. Conservatives typically prefer the world to stay as it is or was, and our world is changing rapidly. Perhaps that makes them unhappy. Obama was president when Wojcik ran his study; perhaps that made liberals happier. There are numerous other possibilities, but the point is that conservatives are claiming to be happier than liberals when the data suggest that they aren't. It's important to keep in mind that conservatives are probably no more likely than liberals to self-deceive, as self-deception is a fundamental aspect of human nature. It's just that conservatives have more to gain from deceiving themselves into believing they're happy than liberals do, and as a consequence they try a little harder to come across as happy when people ask.

Self-Deception Works

These studies suggest that people are highly attuned to the impressions others form of them, but these studies don't indicate whether self-deception is effective. For example, conservatives apparently go through life claiming to be happier than liberals, which means it's safer to ask them how they're doing when you see them on the street, but we don't know whether conservatives benefit from their claims.

The easiest way to answer such a question is to conduct experiments, so that's what we did. Our goal was to test Trivers' hypothesis that self-deception makes people more persuasive, so we broke down the hypothesis into its three logical components. Whenever people are not entirely sure they're telling the truth: first, they should try to convince themselves of the veracity of their claim; second, they should come to believe in their claim; third, by convincing themselves, they should be more effective in convincing others. To test these three possibilities, we borrowed an experimental paradigm from my old friend Peter Ditto.

Ditto's experiments were conducted more than twenty years ago, but they remain my preferred method for examining the effects of self-deceptive motivation on how we gather information. In Ditto's experiments, college students were brought into the laboratory and (incorrectly) told that the investigators were examining the relationship between psychological characteristics and physical health. In service of this ostensible goal, Ditto told the students that they had the opportunity to take a medical test that would indicate whether they were likely to develop a debilitating disease later in life.

In other words, although they were healthy college students now, this test would reveal whether they were likely to become sick in the future. In this "medical test," participants exposed their saliva to a test strip and then observed whether the test strip changed color. Some of the participants were told that color change indicated good news (i.e., they would remain healthy), and some of them were told that color change indicated bad news (i.e., they were likely to become sick in the future).* In reality, the test strip was inert and never changed color.

Imagine yourself in this experiment. You've exposed the test strip to your saliva and are waiting to see if it changes color. If color change means you'll stay healthy, you probably start to worry after a minute has gone by and the strip hasn't changed color yet. What's going on here? Is there something wrong with the test strip? At that point, like many participants in the study, you might throw the strip away and try again.

In contrast, if color change means you're going to get sick, you're probably watching the test strip with a fair bit of trepidation, hoping it sits there doing nothing. Once a half minute or so has passed, you'd probably start to feel relieved, at which point you'd shove the strip into the container for the experimenters, with no plans to check it again later. If that's your intuition about what you'd do, then you've accurately described what most of the participants did in this experiment. Ditto found that people waited longer and rechecked the results more frequently when lack of a color change suggested they were susceptible to future illness than when lack of a color change suggested they would remain healthy.

Although there is nothing wrong with gathering a little more information when you're unsure, *selectively* gathering more or less information depending on whether you like the initial results is a form of self-deception. It's a bit like not bothering to check your

* This study may seem unethical, as Ditto and his colleague were creating unnecessary stress for their participants. It's important to keep in mind, however, that as soon as the experiment ended, Ditto always debriefed the participants about the purpose of the study, and the fact that he had fabricated the disease and the test for it. So, yes, people suffered a bit of stress, but only for the brief interval while they were in the experiment.

exam grade when you're worried you'll get a bad score. Avoiding important information can allow us to deceive ourselves just as it can help us deceive other people (for example, if you change the subject when your spouse asks you why you're late for dinner and you don't want to admit you were chatting with your cute colleague in the next cubicle). Sometimes reality will rear its ugly head; if you failed the exam you'll suffer the consequences whether you know your score or not. But sometimes we can reshape reality by avoiding information. If I make a fool of myself on my blind date but never get in touch with her again, I can approach my next date with more confidence and, consequently, a better shot at success.

In the case of Ditto's experiment, avoiding information is a great way to deceive yourself, because you don't know with certainty what's coming around the bend. If you'd waited longer, maybe the test strip would have changed color, but maybe not, and you were just wasting your time. On average, this sort of selective information-gathering strategy will bias the answers the world gives you in favor of your preferred conclusions, but on any particular day it's possible that the answer you found is the correct one. This sort of selective search can allow people to create a desirable yet potentially distorted view of reality.

Ditto's experiment provides a great procedure for assessing self-deception, but there are a few important pieces missing if we want to test the questions that motivated our research. First and foremost, it's unclear why Ditto's participants tried to deceive themselves about their future. Did they want to protect their happiness and self-esteem from the potentially bad news of their future ill health, as a Freudian account would suggest? Or did they want to preserve an image of themselves as healthy so they could be more confident and hence more effective in attracting romantic partners and allies, as Trivers' account would suggest? In Ditto's case, the answer may well be a bit of both, but of course we wanted to design an experiment that would provide a clear test of

Trivers' interpersonal account. If people engage in self-deception to increase their chances of persuading others, then the type of biased processing showed by Ditto's participants should emerge even when people have no reason to protect their fragile ego.

To test this possibility, my PhD student Megan Smith joined forces with Trivers and me. We conducted an experiment in which participants were told they would see a series of videos in which a guy named Mark engaged in a variety of different behaviors. Megan told some of the participants that they'd be paid a bonus if they could write a particularly persuasive argument that Mark is a dislikeable person, and others that they'd be paid a bonus if they could persuasively argue that Mark is a likeable person (and made sure that all of them knew they had to base their arguments on the information in the videos). The key was that sometimes the initial videos showed Mark engaging in positive behaviors and the later videos showed him engaging in negative behaviors, and sometimes the order of the videos was reversed. Finally, participants were told that they could watch as many or as few videos as they liked, and it was up to them to decide when they were ready to write their persuasive essay.

A few findings emerged from this study. First, rather than watching all the videos and choosing to talk only about the ones that were consistent with their persuasive goals, people gathered information just like Ditto's participants had. If the early videos were consistent with their persuasive goals, they stopped watching pretty quickly. If the early videos were inconsistent with their persuasive goals, they continued to watch for longer, presumably in the hope of finding information that would help them be more persuasive. There is nothing wrong with that, and it's possible that participants knew full well what they were doing and were just trying to be efficient with their time.

To test this possibility, after they wrote their essay, participants were asked for their private opinions of Mark. Their responses indicated that through their biased information gathering they had convinced themselves that Mark was what they wanted him to be: people who were paid to argue in his favor found him more likeable than people who were paid to argue against him. When we offered them another bonus if they could accurately guess how others would feel, they thought their own impressions would be shared by others. Again, this result suggests they were unaware of the impact of their own biased information gathering.

Finally, we found that the most persuasive people were the ones who were biased in their information gathering; participants who were unbiased wrote arguments that other people found less convincing. This effect of biased information gathering emerged primarily to the degree that the participants convinced themselves. When their biased information gathering enabled them to see Mark as they were paid to portray him, their arguments were most effective.

This is just a single experiment, but it provides support for Trivers' proposal that we deceive ourselves in order to deceive others more effectively. The study also helps throw some light on a variety of behaviors that we see in our everyday lives. For example, there was a fair bit of concern after the 2016 presidential election that "fake news" might have played a role in Trump's victory. Although this is certainly possible, the results of our study suggest that people are biased toward gathering information that fits with what they want to believe. Consistent with this interpretation, research conducted after the election showed that the people who were consuming most of the fake news were partisans who already had strong beliefs in the direction of the fake news stories.

This finding suggests that fake news probably had very little effect in causing undecided voters to choose Trump, but it was probably very persuasive to people who already tended to support Trump. In other words, people sought out and believed fake news only if it was consistent with their prior beliefs, and thus the effect of fake news was likely just greater polarization of the electorate. I suspect that only in rare cases did fake news actually generate political preferences that didn't previously exist. Such a possibility would also be consistent with Chip Heath's research on urban legends discussed earlier: people pass along exaggerated stories in part because exaggeration ensures that others share the teller's emotional reaction. In this sense, fake news probably serves a bonding function, bringing group members closer together in their outrage at the supposed misdeeds of the opposing party.

It might strike you as odd that people are susceptible to these sorts of biases, but it's worth remembering the evolutionary pressures that made us so smart in the first place. As the social brain hypothesis proposes, a substantial reason we evolved such large brains is to navigate our social world. In contrast to value in the physical world, value in the social world reflects objective reality only partially. If we decide that bell-bottoms are cool, then cool is what they are, and you'd better get yourself a pair or risk being a wallflower at the disco. A great deal of the value that exists in the social world is created by consensus rather than discovered in an objective sense.

If I can influence the consensus to move in a direction that favors me (whatever I'm doing is cool), then I'll probably benefit even if my objective understanding of the world is biased. For this reason, it makes sense that our cognitive machinery evolved to be only partially constrained by objective reality, as the social consequences of our beliefs are often just as important as the objective consequences. Indeed, some researchers have argued that our minds evolved the ability to process logical arguments not so we could discover the true state of the world, but so we could convince others of the accuracy of our own self-serving beliefs.

In this sense, the social brain hypothesis suggests that the great discoveries of humankind are really just an evolutionary by-product of our ancestors' efforts to persuade others of their dubious claims.

In the words of my astronomer brother, "So NASA can thank all the self-serving liars of our evolutionary past for our ability to send robotic spacecraft throughout the solar system?" The answer to this question is a resounding yes, and it speaks to just how important sociality is in the evolution of our incredible cognitive abilities.

6

Homo Innovatio

Innovation of new products is not a uniquely human activity. Chimpanzees strip leaves off branches to fish for termites, and New Caledonian crows fashion hooks out of palm fronds to extract insects from holes in dead trees. These are wonderful examples of animal ingenuity, but even the most cursory glance at our own world indicates that we're operating on a different level. We create comfortable homes for ourselves in every environment on earth; we harvest, store, and deliver a wide variety of foods to eat; we communicate with one another instantaneously across vast reaches of the globe; and we entertain ourselves with all sorts of complex gadgets.

Many of our inventions are so new that they didn't exist a generation ago, but our lives look very different from those of other animals on this planet whether you go back one hundred, one thousand, ten thousand, or even one hundred thousand years. In all these time periods, humans protected themselves from predators and the elements, and preyed on much larger and stronger animals, using clothing, shelter, tools, and strategies they invented.

The innovations of our ancestors and peers permeate every aspect of our lives. Their inventions enable me to tell stories to people I've never met by tapping at my keyboard, while taking the occasional break to make a meal or a cup of coffee, all without leaving my well-lit and climate-controlled home. Meanwhile, halfway around the globe, my chimp cousins still sit on tree branches, in the hot sun, driving rain, and chilly nights, making a living with their bare hands just as they did when our ancestors bid them farewell six million years ago. The single feature that most notably differentiates our position from theirs is our inventiveness. And herein lies the paradox: technical innovation is the defining feature of our species, but most people never invent anything.

When innovation researchers ask representative samples of people whether they have modified any products at home or created anything new from scratch (such as tools, toys, sporting equipment, cars, or household equipment), about 5 percent report that they have done so in the last three years.* The percentage of innovators varies a bit by country, but never cracks 10 percent. For such an innovative species, one in ten or twenty seems awfully low. Yet, when I reflect on my own life, I can't recall ever inventing anything. I have a few inventive friends, but I'd be surprised if 5 percent of them have ever invented anything either, let alone in the last three years.

These numbers suggest an extraordinary disconnect between *Homo sapiens* as a species and individual humans. When you think

*This might seem like an odd question to ask of the general public, but a surprising percentage of important inventions have been created by people modifying objects for their own use. One of my favorite examples is from 1911, when Ray Harroun decided to drive alone in the first Indianapolis 500 rather than with a racing mechanic. This decision required him to devise a way to monitor the cars behind him, which was traditionally the job of the mechanic. To achieve this goal, he installed what is believed to be the first rearview mirror in an automobile, thereby assuring race officials that he could safely race solo. (The Indianapolis 500 was run on a brick surface at the time, and Harroun later admitted that he couldn't see a thing in his jiggling rearview mirror, but he won the race anyway.) A more mundane example of user innovation can be found in the case of the skateboard, which was invented in the 1940s by numerous kids across the United States and Europe who took apart their roller skates and nailed the wheel mounts onto crates or boards.

of other species, their defining qualities are shared by all their members. Elephants are huge and strong, and "huge and strong" pretty much describes every elephant I've ever seen. Cheetahs are fast, lions and tigers are fierce, and dolphins are playful, and those adjectives pretty much cover all of them as well.

There are at least three ways to interpret this disconnect in inventiveness between *Homo sapiens* and individual humans: first, most of us aren't inventive; second, all the obvious inventions were thought of long ago; and third, most people are inventive but don't direct their innovative tendencies toward making new stuff.

To start with the first possibility, perhaps most people are ill-suited to innovate and only the rare geniuses among us have the capacity to invent new things. Extraordinary innovations such as the telephone, lightbulb, and jet engine are consistent with such a possibility, as the insights underlying them seem out of reach for ordinary minds. According to this possibility, technical innovations are analogous to genetic mutations; they are rare and mostly trivial or even worthless, but the occasional product sweeps through the population and has an enormous impact on the species. If that's the case, then the disconnect between our species and its individual members is a fundamental one. According to this possibility, most humans are not innovative at all; we just have the good sense to benefit from the rare geniuses among us who invent things that improve our lives.

Alternatively, perhaps most people *are* innovative, but all the obvious and simple products have already been invented. Maybe as recently as a few hundred years ago pretty much anyone could invent stuff that had a good chance of being useful. According to this possibility, we happen to live in a unique window of time in which inventions have become so complex that they are now limited to geniuses and large teams of techies. Super-complicated inventions such as the iPhone would seem to support such a possibility, as there are literally hundreds of patents that underlie this one tool.

This view of human invention is a common one, and people regularly suggest it when I talk about innovation. I have a three-word response: *wheels on suitcases.*

At least since the steamship, people have traveled the globe with relative ease and regularity. Yet, across all these generations of travelers, no one thought to put wheels on suitcases until 1970, and they didn't catch on until the modern version of a wheeled suitcase with a retractable handle appeared in 1987.* This failure to attach wheels to suitcases was all the more remarkable given that once people lugged their nonwheeled suitcases to the airport, they then paid cold, hard cash to a porter who plunked their nonwheeled suitcases on his cart and easily *wheeled* a whole family's worth of baggage the last fifty yards to the ticket counter.

If you haven't traveled with nonwheeled suitcases, you have no idea what a nuisance it is. In the early 1980s, I traveled across the country to college, which meant that I had to carry two large and heavy suitcases back and forth at the beginning and end of every summer. Because I'm not the tallest human on the planet, my suitcases would drag on the ground unless I hunched up my shoulders and bent my elbows to keep them aloft. It looks incredibly pathetic to drag your nonwheeled suitcase on the ground (and it's hard on the suitcase), so this meant that I was always hurrying across massive airport halls with hunched shoulders, bent elbows, and hard suitcases banging into my shins and knees. By the time I got to the ticket counter, I had a stiff neck, my legs were bruised and battered, and I was a sweaty mess.

It's hard to imagine a more fitting start to a transcontinental

* Maybe wheeled suitcases wouldn't have been very useful until the last hundred years or so, as people would have been wheeling their suitcases through the mud or across bumpy cobblestones. Since at least the 1870s, however, wheeled suitcases would have been very handy at the great train stations of the world, such as those in New York, London, and Berlin. By this accounting, it still took at least one hundred years for someone to come up with this simple innovation.

flight than that. But I was a college student, and paying a porter meant one less pizza when I arrived at the dorm, so like almost everyone else, I suffered through it. Not once did it occur to me that I had an opportunity to invent something new and incredibly simple that would make my life easier and earn me a fortune. I don't have the insight to predict when the next incredibly simple invention will come along that will vastly improve some aspect of our lives, but I can promise you that it will happen again.*

The luggage example brings us to our third possibility: that most people are capable of innovating but are disinclined to turn their efforts toward product innovation. According to this possibility, technical innovation doesn't necessarily demand genius. After all, wheeled suitcases aren't exactly rocket science. Rather, technical innovation is rare only because most people focus their energies elsewhere. Where is elsewhere? By way of answering this question, let's consider an extraordinary sequence of events witnessed by Jane Goodall while she was observing chimps in Gombe, Tanzania. Forgive me for the fact that the story I'm about to relate is an unpleasant one. When I read it, I couldn't put it out of my mind for days.

The brief background to this story is as follows: Melissa is a chimp who has just had a new baby. Passion is another chimp in Melissa's group. Pom is Passion's adolescent daughter, and Passion and Pom are vicious psychopaths. Here is what Goodall wrote, in slightly abbreviated form:

> At 17:10 Melissa, with her three-week-old female infant, Genie, climbed to a low branch of a tree. Passion and her daugh-

* For example, a quick Google search yields all sorts of clever and simple inventions: a baby stroller/scooter hybrid; a citrus spritzer that is just a spray nozzle with a sharpened base that you jam into a lemon or lime; a pair of pizza scissors with a spatula at the base to catch and deliver your slice; or a fork with wiggly tines so the twirled spaghetti doesn't fall off so easily.

> ter Pom cooperated in the attack; as Passion held Melissa to the ground, biting at her face and hands, Pom tried to pull away the infant. Melissa, ignoring this savaging, struggled with Pom. Passion then grabbed one of Melissa's hands and bit the fingers repeatedly, chewing on them. Simultaneously Pom, reaching into Melissa's lap, managed to bite the head of the baby. Then, using one foot, Passion pushed at Melissa's chest while Pom pulled at her hands.
>
> Finally, Pom managed to run off with the infant and climb a tree. Melissa tried to climb also but fell back. She watched from the ground as Passion took the body and began to feed. Fifteen minutes after the loss of her infant, Melissa approached Passion. The two mothers stared at each other; then Melissa reached out and Passion touched her bleeding hand. As Passion continued to feed on the infant, Melissa began to dab her (own) wounds. Her face was badly swollen, her hands lacerated, her rump bleeding heavily. At 18:30 Melissa again reached Passion, and the two females briefly held hands.

What bothered me most about this story was not the cannibalism itself, disturbing though it is, but the fact that Melissa reconciled so quickly with the two killers. Worse yet, this wasn't an isolated incident. Passion and Pom continued to kill and eat newborn infants in their group for years. One poor mother lost three babies in a row, and it was then that Goodall realized that only one infant had survived its first month in the group in the last three years. Despite the simplicity and predictability of Passion and Pom's attacks, none of the mothers devised a successful strategy for dealing with this pair of cannibals, and the mother-daughter team devastated the reproductive potential of their group. The other mothers responded much as Melissa had, fighting their hardest during the attack but then seemingly accepting their fate and doing nothing about it. It was their helplessness in the face of such a terrible but solvable problem that haunted my thoughts.

There is no way to know what Melissa was thinking when she held out her hand to Passion while her baby was still being consumed. My guess is that she was stymied by the fact that she could devise no way to defeat Passion and Pom and, lacking any alternative, she thought it best to reconcile. Chimps are awfully clever, but there are strict limits to their cognitive capacities. In many ways, their problem-solving abilities are similar to those of a human toddler. Chimps can and do plan, such as when they prepare a termite fishing stick prior to going to a termite mound, but they can't create and mentally test complex scenarios that could lead to a variety of outcomes. Rather, the only way for a chimp to test a plan is to physically enact it and see what happens. They are incapable of turning over complex problems in their minds.

Imagination and simulation are some of the major advantages provided by our big brains.* Life is full of problems that are dangerous to solve without thinking them through in advance. Once you've started to enact a plan, it's often impossible to stop and start over. If I discover halfway through my plan to kill Passion and Pom while they're sleeping that it's not working because one of them woke up while I was attacking the other, I can hardly ask them to go back to sleep while I go back to the drawing board.

* In his superb book *Sapiens*, Yuval Harari argues that our capacity to create elaborate fictions accounts for our propensity to cooperate with each other well beyond the groups in which we originally evolved. In Harari's words, "If you tried to bunch together thousands of chimpanzees into Tiananmen Square, Wall Street, the Vatican or the headquarters of the United Nations, the result would be pandemonium. By contrast, Sapiens regularly gather by the thousands in such places. Together, they create orderly patterns—such as trade networks, mass celebrations, and political institutions—that they could never have created in isolation. The real difference between us and chimpanzees is the mythical glue that binds together large numbers of individuals, families, and groups. This glue has made us the masters of creation."

Our large brains have cracked this particular nut via our capacity to build and mentally simulate complex plans that account for numerous contingencies. In the case of Passion and Pom, you need only reflect on how you would respond if someone attacked you and ate your baby to realize that solving this problem is well within the scope of every one of us. If I lived in a world without police or government and Passion and Pom attacked me, I suspect my first impulse would be to attack them back. Knowing that this would be a hopeless battle, however, I would control my initial impulse and would pretend to reconcile while I developed a plan to neutralize these psychopaths.

Perhaps I would attack them in their sleep, or maybe I'd solicit the help of friends who had been victimized by them. The point is not what I would actually do to solve the problem. Rather, the point is that we all have the capacity to make a plan, simulate it mentally to see if it would work, find points of weakness in the plan through this simulation, update and improve it, and continue to iterate through this process until we are satisfied with our approach. Furthermore, we can do all this while lying in bed, sitting by the fire, or driving to work—there is no physical evidence that we are engaged in the process, so no one knows whether we're busy plotting or just daydreaming. But because such mental processes exceed the capabilities of chimpanzees, the chimp mothers were all helpless in the face of these two deranged cannibals.

In contrast to chimpanzees, hunter-gatherers in every society on earth are adept at dealing with difficult group members. These problems are universally addressed via social strategies such as communal ostracism or coalitional violence against the offending individuals. When the strongest or most aggressive members of hunter-gatherer communities cause problems, their peers typically band together in response. First, they mock the bullies, which serves as a warning that the behavior is unacceptable. If mocking doesn't bring them in line, the group often pretends that the bullies aren't there, ignoring them when they speak and talking about them as if they were absent. If that soft form of communal ostracism doesn't work, then one morning the bully wakes up to find himself alone, as everyone else has packed up in the middle of the night and left. If that doesn't work, one morning the bully doesn't wake up at all.

By banding together, our ancestors found that even the strongest and most aggressive individuals in their group were no match for the rest of them. How does that fact explain the disconnect between the inventiveness of our species and its individual members? And what does all this have to do with wheeled suitcases and why we took so long to invent them? The answer to both questions is that our ancestors were *social innovators*. That is, they solved their problems socially rather than by inventing new products, and so do we. Why? Because the sociality we evolved after we moved to the savannah leads us to direct our incredible problem-solving capacities toward social rather than technical solutions.

Most people find social relationships and social reciprocity rewarding, and as a result they gravitate toward social solutions to their problems even if they are capable of creating a technical solution. This does not mean that we evolved innovation capacities that are restricted to the social domain. The key abilities that underlie innovation, such as scenario building and mental simulation, are useful in a wide variety of domains and can be used to solve social or technical problems. The issue is where we tend to direct those innovative abilities.

If we evolved to innovate socially, this possibility could help explain the extraordinary disconnect in technical innovation between *Homo sapiens* as a species and individual humans. Technical innovation may be rare because people are preoccupied with the search for social solutions to their problems. If we return to the example of wheeled suitcases, history suggests that our default social orientation (i.e., locate a friend or porter to help us) was blocking our capacity to find a glaringly obvious technical solution (i.e., put wheels on the suitcase in the first place).

Indeed, the competition between social and technical thinking appears to be fundamental to the way our brain itself functions. In *Social: Why Our Brains Are Wired to Connect*, Matthew Lieberman argues that the default, or "resting," state of our brain is not one of rest at all, but of perpetual activation of the social neural network. We start to activate this network when we are newborns, before we know anything about the social world, and this network continues to capture our thoughts throughout our adult life. Furthermore, activation of the social network is associated with deactivation of nonsocial thinking, and vice versa. As a consequence, when we are freely pondering our world, our social orientation crowds out other problem-solving approaches.

In my own case, whenever I went to the airport laden with luggage, my first thought was whether I could talk a family member or friend into helping out. Barring that, my next thought was whether it was worth hiring a porter. Because these solutions were always at the front of my mind, I (and every other traveler) never considered wheels as an obvious alternative solution. When a luggage manufacturer put wheels on suitcases in the 1970s, and a pilot improved the design in the late '80s with better wheels and a retractable handle, a very simple technical solution finally pushed its way past our social orientation and into our collective consciousness. The moment I saw a wheeled suitcase for the first time, I knew I needed one, even though it had never occurred to me to put wheels on my own suitcase.

What Are Social Innovations?

Before considering the evidence in support of this social innovation hypothesis, we need to sort out what it means for an innovation to be social, beyond the obvious fact that it involves people.

An innovation is a new way to solve a problem—at least, it's new to the solver if not to the world—and thus a social innovation is a new way to solve a problem through the use of social relationships. The key issue is not the nature of the problem, but the nature of the solution.

For example, the desire to talk to faraway friends and relatives is a social problem, and it can be solved technically, by inventing a telephone, or socially, by passing messages through friends. Similarly, the desire to walk with a broken knee is a technical problem, and it can be solved socially by having your friends assist you or technically through the aid of crutches. When these solutions are novel, they are social or technical innovations; when these solutions are repeated or borrowed from others, they are social or technical solutions but not innovations. Asking for help is a social solution, but it's not a social innovation (unless it's the first time it ever occurred to you to ask for help).

If you consider the greatest inventions in human history, it becomes obvious that many of them are social. For example, one of *Homo erectus'* most important inventions was division of labor and the resultant social coordination. By dividing up tasks and working together as a group, *Homo erectus* were able to hunt massive animals with the simplest of tools. Division of labor made groups more than just the sum of their parts, and thereby played an outsize role in making us the success story we are today.

Despite our much briefer time on the planet, *Homo sapiens* have created exponentially more technical inventions than *Homo erectus* did, and we have devised countless social innovations as well. For example, although love of money may be the root of all kinds of evil, money is an incredibly useful social innovation. The physical object itself is trivial—what is important about money is the social convention by which everyone agrees on its value. After division of labor, money may be the second most important social innovation in history. Money can be used to acquire anything of value, and thereby enables all sorts of market exchanges that would be nearly impossible if people were required to trade in goods. Imagine the inconvenience of going to the mall to buy a new sweater and bringing along a pig or goat with you as barter. In the world before money, that's how people shopped.

If division of labor is our most important social innovation, and money comes in second place, what's next? My choice for third place might sound odd, but I'd be inclined to put waiting in line next. I never gave much thought to the social innovation of waiting in line until I found myself in a situation in which lines didn't exist. I was trying to cross an international border that allowed only foot traffic and I was coming from a rural area that apparently had no conventions about waiting in line. There were several dozen fellow travelers trying to get their passports stamped by one person in a tiny booth, and not a single one of them waited in line. As I surveyed the swirling mass of humanity enveloping the customs booth, I was reminded of the BBC special *Planet Earth*, in which emperor penguins huddled together in Antarctica in a giant penguin blob in their effort to stay warm.

Contemplating this human blob, I could feel my heart sink in my chest, and I wondered if I should change my travel plans. But I had to cross the border, so I chose a random spot in the blob and joined in, trying to edge closer to the booth whenever possible. A couple of times I came heartbreakingly close, only to have the current change direction at the last minute. Sometimes I moved forward with surprising speed, and sometimes I moved backward.

After what seemed like an eternity squeezed into this sweaty and dusty mass of my fellow travelers, and just as I was wondering if I should (or even could) extricate myself from the blob and try again the next day, an eddy in the human current plunked me right in front of the booth. I thrust my passport at the harried customs officer, and it seemed like my lucky break, until the crowd started carrying me off while he still had my passport. Out of desperation,

I grabbed the plywood counter with both hands. The customs officer slipped my stamped passport between my fingers just before my feet left the ground and I lost my grip. Since then, I have regarded the practice of waiting in line as one of the world's great social innovations.

Money and lines are ancient inventions, but the internet has recently enabled all sorts of social innovations that our ancestors would have loved. Everyone has a favorite, but I regard social media and dating sites as some of the most important social innovations on the internet.* When I was young, there were basically three ways to meet your life partner. You could hang out in places where your partner was likely to be found, you could be set up by mutual friends, or you could put a "personal ad" in the back of your local newspaper, in which you listed your height, weight, and age, and provided a four- or five-word blurb about why someone should choose you. Newspaper photos were expensive and of poor quality, so people simply claimed to be attractive and didn't bother to include photos in these ads. My guess is that the hit rate for this method of finding a partner was incredibly low, and it certainly didn't have a reputation as a great way to meet Ms. or Mr. Right.

Social networking and dating sites have moved us a million miles beyond that world, by offering specialized places for people with particular interests or backgrounds, by allowing people to learn a lot more about each other before they bother to meet in person, and most notably by massively increasing the number of potential partners people can meet. If dating is a numbers game, and I suspect it is, the internet plays a critical role by helping peo-

* Keep in mind that the distinction between social and technical innovations is continuous rather than dichotomous. For example, Facebook is an incredible social innovation, but it's also a remarkable technical achievement. Thus, it's closer to the middle of the social–technical continuum than waiting in line (a purely social solution to the problem of who gets to go next) or the lightbulb (a purely technical solution to the problem of unwanted darkness).

ple find their needle in the haystack. Social networking and dating sites get a ton of traffic, and research shows that romantic partners are increasingly likely to meet online.

For example, a study of nearly 20,000 people who married between 2005 and 2012 by John Cacioppo of the University of Chicago and his colleagues found that over one-third of them originally met online. Although the study concluded in 2012, 7.7 percent of those who'd met in person had already separated or divorced, while only 6 percent of those who'd met online had separated or divorced. That might seem like a small difference, but every percentage point in the offline sample is equivalent to another 125 broken relationships.

Furthermore, if we compare the 4,000 people who met on social network and dating sites (the two most common online meeting sites) to the 8,000 people who met at work, through friends, at school, at a social gathering, or at a bar or club (the five most common offline meetings points), we see reliable differences in marital satisfaction. None of these samples who met offline was as satisfied as the online samples, although those who met at school or a social gathering came close. Couples who met through friends or at a bar, in contrast, were least likely to be satisfied in their marriage. These data suggest that relationships that begin online may be more likely to last than relationships that begin almost anywhere else.

To turn to social media more broadly, there are numerous instances of the value of Facebook, YouTube, and Snapchat. For example, although the Arab Spring hasn't worked out very well (yet), Facebook played a critical role in helping relatively powerless individuals coordinate massive protests against totalitarian regimes across the Middle East. Facebook may be ubiquitous and occasionally of great societal importance, and it has helped me reconnect with friends from high school, but my favorite example of the power of social media is YouTube.

When I was a kid, gatekeepers held the keys to almost every

route to fame and fortune. If you wanted to be a movie star, someone had to decide that you were star material. If you had an idea for a new type of TV show, someone had to decide that your idea was entertaining. YouTube eliminated these gatekeepers in one fell swoop—now anyone with a computer and an idea can start putting up videos for the world to see. YouTube has demonstrated that gatekeepers have a very limited sense of what people really want. The most compelling evidence I know for this claim is the incredible popularity of YouTubers who film themselves playing video games. It never would have occurred to me that people would want to watch other people play video games, but one afternoon I heard yelling from my son's room and asked if he was all right. It turned out the yelling was coming from the YouTube channel he was watching, in which people were playing video games and yelling at their screens.

In case you think this is a fringe market, as of this writing PewDiePie is the king of this genre, with more than 54 million subscribers. To put that number in perspective, the most popular TV show in America in the 2015–16 season was *Sunday Night Football*, with an average weekly audience of 22 million. When I googled PewDiePie's audience over the last month, he had 147 million views, substantially more than the 90 million views of *Sunday Night Football* each month. Almost every video he puts up has millions of viewers, who make him a fortune from advertisements (and his videos are a lot cheaper to produce than a football game).

PewDiePie is far from the only success story on YouTube. Various other YouTubers make a living demonstrating different techniques for putting on makeup, kidding around about bizarre conspiracy theories, offering life hacks, and even just showing scenes from their daily lives. And there are no age barriers, as a few small children make huge sums of money just opening presents and playing with them. What I love about the most successful channels on YouTube is how unappealing they are to most adults. None of the most successful YouTubers would ever get past the security guy at a talent agency, much less an actual talent agent, but they clearly resonate with their target audience. Social media has democratized the route to fame and fortune, and in so doing, has filled vast unmet entertainment needs (e.g., to watch people pop their pimples or lovingly unbox sneakers).

The Social Innovation Hypothesis

So how do we explain why some people innovate new products but most people don't? If humans evolved to solve their problems socially rather than technically, that doesn't mean that everyone will choose a social solution every time. Rather, we should be able to use this hypothesis as a starting point to predict who will innovate new products and who won't.

An obvious place to begin such a search is with the prediction that people who are less social should be more likely to innovate technically. Not only do less-social people have smaller social networks and thus fewer people to turn to for assistance, but many of them also find social solutions less rewarding and reliable. As a consequence, less-social people should be likely to orient themselves toward technical solutions to their problems, a proclivity that would lead to a greater rate of technical innovation.

If we conceive of social and technical orientations as somewhat unrelated traits, then we would expect people to lie in one of four quadrants, defined by their technical and social orientations (see Figure 6.1). Given how critical social functioning was to our ancestors' success, most people are likely to be found in the upper two quadrants of this diagram. Those on the upper left have relatively weak technical skills, so we would expect them to innovate socially. But the key point is that those on the upper right, who have strong technical skills, should also innovate socially rather than technically.

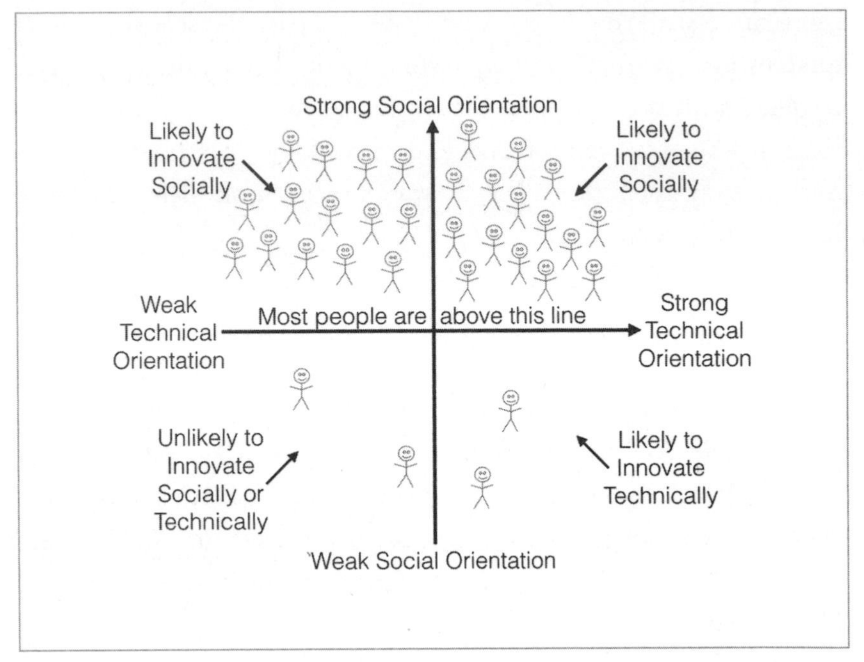

Figure 6.1. Social orientation, technical orientation, and innovation. (Adapted from von Hippel and Suddendorf, in press)

For most of us, social engagement is fun and rewarding, and ever since we left the trees, it has been our default orientation toward the world.* As a consequence, highly social people who have the talent to tinker with objects are relatively unlikely to do so when confronted with a problem. For example, it might be easy enough for some people to wire a can opener to their toaster and an oven timer to create an automatic dog feeder for when they're out of town, but even people as handy as that would typically prefer to ask a friend to come over to feed Fido while they're away. Thus, the social innovation hypothesis predicts that most people are unlikely to innovate technically. Of the (relatively few) people who reside in the lower two quadrants, only those who have a strong technical orientation

* This observation holds for introverts, too. Almost everyone likes to socialize; introverts just prefer to socialize less often and with a smaller circle of closer friends.

would be likely to innovate technically. So even if technical skills themselves are relatively common, the ubiquity of human sociality would make technical innovation a relatively rare occurrence.

It is very difficult to test these possibilities with ancestral data, but two pieces of modern evidence support the social innovation hypothesis. First, one way of quantifying sociality is to look at the frequency of autism. People on the autism spectrum vary in intelligence, but regardless of their intellect, they struggle with social relations. Impaired social functioning is one of the hallmarks of autism. Even highly intelligent individuals with autism have problems with Theory of Mind, as their brains don't automatically compute the intentions and feelings of others in the manner we discuss in chapter 2. As a result, people with autism don't understand neurotypical people very well and struggle to engage them socially.

Given these facts, it's no surprise that you rarely find people with autism working in sales, and they are also rare in the humanities and social sciences. In contrast, people on the autism spectrum can be readily found in fields in which the dominant orientation is toward objects and away from people, such as engineering and the physical sciences. For example, Simon Baron-Cohen of Cambridge University and his colleagues found that autism is more common in the families of physicists, engineers, and mathematicians than in the population in general.

When Baron-Cohen and his colleagues went on to develop a scale to quantify levels of autism, one of their first comparisons was between students in the sciences and those in the humanities. They found that science students had higher autism scores, including reduced levels of sociality, than humanities students, and this difference was most notable among students studying physical sciences, computer science, and mathematics. Students in the social sciences scored no differently from students in the humanities. Engineering students in this sample had scores that fell between those in the physical sciences and those in the humanities.

Unsurprisingly, engineers and physical scientists are also more likely than people in the humanities and social sciences to hold patents or to innovate technical products for their own use at home. In other words, among engineers and physical scientists, there is a preponderance of people in the bottom right-hand quadrant of Figure 6.1. As a notable example, Silicon Valley is a hotbed of innovation and also features an unusual concentration of people on the autism spectrum.* Of course, such findings are not evidence that sociality is *preventing* people from innovating technically (as proposed by the social innovation hypothesis), but the fact that sociality and technical innovation appear negatively correlated raises the possibility that one influences the other.

As a second approach, we can look for sex differences in innovation. It's always dicey business investigating differences between the sexes, given the widespread tendency on both sides of the aisle to extrapolate well beyond the data. But please bear with me through this discussion, and I think you'll see that the implications make perfect sense. If we take the plunge into this literature despite the obvious risks, we find that one of the most substantial and widely replicated sex differences in preferences is the tendency for men to be more interested in objects and women to be more interested in people. For example, this result was confirmed in an analysis of vocational interests of more than half a million men and women from the United States and Canada.

No doubt such sex differences in interests are partially a function of sex-typed cultural expectations, but research suggests that

* Autism rates fluctuate by time and location (an important but poorly understood fact in and of itself), and it is no easy matter to pin down the exact rate of autism in the large and somewhat amorphous Silicon Valley. But a study by Baron-Cohen and his colleagues compared the "Silicon Valley of the Netherlands" (Eindhoven, where 30 percent of the jobs are in the IT sector) to two similarly sized Dutch cities (Utrecht and Haarlem). They found that autism rates were 23 per 1,000 people in Eindhoven, but only 6 per 1,000 in Utrecht and 8 per 1,000 in Haarlem.

they are also partly innate and appear to be culturally universal. Many studies document sex-typical differences in toy preferences that are evident in infants in their first year of life (e.g., trucks versus dolls), and the male preference for toy trucks has even been shown in monkeys.

Regardless of whether these differences are driven by cultural expectations or biological differences (or, more likely, both), the social innovation hypothesis suggests that sex differences in sociality are likely to lead to sex differences in technical innovation. Consistent with this possibility, professions oriented toward technical solutions (e.g., mathematics and engineering) boast many more people in their ranks who not only are on the autism spectrum but are predominantly male.

An enormous body of cross-cultural research shows that women typically have stronger verbal than spatial skills and men typically have stronger spatial than verbal skills, so the fact that men are overrepresented in math and engineering might reflect nothing more than sex differences in ability profiles. We are naturally drawn to things we're good at, and we tend to improve in domains that interest us, so the causal order of these sex differences in abilities and interests is hard to pin down. Women are more interested in people, so perhaps they communicate with one another more and become more verbally skilled. Men are more interested in objects, so perhaps they spend more time manipulating them, which fosters better spatial reasoning. Or perhaps the causal order goes the other way, or both ways.* The key finding, however, is that even among men and women who have strong mathematical and technical skills, men show greater interest in objects and women show greater interest in people.

For example, David Lubinski of Vanderbilt University and his

* Whatever the origins, it's important to keep in mind that male and female spatial and verbal abilities overlap more than they differ.

colleagues selected a sample of over fifteen hundred mathematically precocious students across the United States and followed them into adulthood. By the time they had reached middle adulthood, men from this sample were more than twice as likely as women to hold a patent. More important for our purposes, sex differences in social interests matched the sex differences in technical innovation.

When asked about their work preferences, the men in this sample were more interested than the women in "Working with things (e.g., computers, tools)" and "Inventing/creating something with impact." In contrast, the women were more interested than the men in "Working with other people" and "Having the results of my work significantly affect other people." Similarly, the women were more interested than the men in "Time to socialize" and "Strong friendships." Thus, even among mathematically gifted men and women, notable sex differences emerge in their social orientation and technical achievement.

These sex differences in social orientation also predict whether people choose a career that involves technical innovation. The clearest example of such an effect can be seen in a longitudinal study by Ming-Te Wang of the University of Pittsburgh and his colleagues with another national sample of approximately fifteen hundred students. Based on the students' SAT scores, Wang separated students who had high math *and* high verbal abilities from those who had high math but only moderate verbal abilities. Several important differences emerged between these two groups of students, despite the fact that they were all great at math.

First, Wang found that the high-math/high-verbal group was two-thirds female, whereas the high-math/moderate-verbal group was two-thirds male. In other words, women who are mathematically gifted tend to be smart across the board, whereas a lot of men are great at math/science and not much else. This turns out to be important because people who are gifted mathematically and verbally reported greater interest in working with people and less interest in working with objects than those who were great only at math.

Wang also found that students who were great only at math were much more likely to go into a career in the physical sciences and engineering than students who were smart across the board. These career choices matched their earlier stated interests; the more they were interested in working with people, the less likely they were to choose a career in physical science, and the more they were interested in working with things, the more likely they were to choose a career in physical science.

These findings are important for several reasons. First, they suggest that women's underrepresentation in math and the physical sciences might not be a problem in the classic sense that it signifies barriers against women in such fields. Many people have interpreted female underrepresentation in math and science as evidence that women are made to feel unwelcome in those fields. There is evidence on both sides of this issue, but Wang's data suggest that gender stereotypes and a potentially hostile climate are not the primary factors keeping most women out of these fields. After all, women used to be rare in the biological sciences as well, which they now dominate at both the undergraduate and graduate levels. Presumably the climate was no more welcoming to women in biology (or the numerous other fields in which female participation soared) than it was in math, engineering, and the physical sciences (where female participation has risen much more slowly).

Rather, Wang's findings suggest that most people who are talented verbally as well as mathematically are not terribly interested in becoming physical scientists and engineers. Because women who are good at math tend to be smart across the board, they vote with their feet and choose other careers. This result makes perfect sense to me, as it reminds me of my own career choice. Like one-third of the men in Wang's study, I have a female-typical brain in that my verbal abilities are stronger than my math abilities. When I was in high school, I considered becoming an engineer, and even took the engineering entrance exams for college before I realized that math and building things didn't interest me as much as a people-oriented career in the social sciences.

It's important to remember that I had the luxury of making this choice because I grew up in a wealthy country where people can make a perfectly good living with a liberal arts education.* In contrast, in poorer countries where most of the good jobs are in technology and science, people who are good in math and science tend to follow this career path. This difference in career choices that emerges as a function of the overall wealth of a country leads to an interesting and counterintuitive effect. Poorer countries are typically less gender-egalitarian than wealthier countries, but women are more likely to enter the sciences in these poorer, less gender-egalitarian countries than they are in richer and more egalitarian ones. If sexism or cultural mores about what women should do were keeping women out of science as was once the case, this result is the opposite of what we would expect. In contrast, if underrepresentation of women in math and science is now largely a function of gender differences in interest leading to gender differences in career choice, then this result is exactly what we would expect. Engineering, math, and the physical sciences may simply not be very interesting to the majority of people who are socially oriented and have other options.†

* Notwithstanding the old joke: How do you get a psychology graduate off your porch? Pay for the pizza.

† Indeed, it's possible that sex differences in college graduation rates will prove to be a much bigger problem than sex differences in math and science participation. Since the late twentieth century, women have been more likely than men to graduate from college in the United States. On the one hand, you could view such statistics as righting a historical wrong, given that women were once prevented from attending some of the best colleges. On the other hand, women rarely marry less-educated men, so differential graduation rates may cause relationship problems for the foreseeable future.

Because women tend to be oriented more toward people and less toward objects, the social innovation hypothesis suggests that women's innovations will be less likely than men's to be targeted at technical solutions. The data are largely consistent with this possibility. For example, in a sample of nearly ten thousand patents granted in six European countries in the 1990s, less than 3 percent were held by women. Cultural and historical factors that disadvantage female inventors assuredly play a role in this figure, but it is noteworthy that the percentage of female patents in that sample is four times lower than the percentage of female engineers (approximately 12 percent).

Similarly, in a representative sample of nearly twelve hundred people from the United Kingdom, 8.6 percent of males and 3.7 percent of females reported that they engaged in technical innovation by creating or modifying products for their own purposes in their own home. These data suggest that even when the formal constraints, biases, and barriers involved in patenting are removed, the ratio of male to female technical innovators is still greater than two to one. Even though technical innovation is rare for both men and women, men innovate new products more often than women do, both formally and informally. According to the social innovation hypothesis, the fact that men are overrepresented among inventors is not evidence that women are less inventive, but rather that women's sociality leads them to innovate elsewhere.*

Contrary to the widespread belief that inventors are a rare breed, the social innovation hypothesis clarifies that almost all humans *are* innovators, but most people direct their inventive capacities toward social rather than technical solutions. Humans

---

Because marriage is good for men, and unmarried men are bad for society, this sex difference in education may lead to all sorts of individual and societal disruptions.

* Because there is virtually no research on social innovation, the possibility that women are more socially innovative than men remains an untested hypothesis as of this writing.

don't innovate every day, but that's only because we don't have to—our shared cultural knowledge provides ready-made solutions to most of the problems we face. However, we are capable of innovating whenever we encounter a novel problem that is of sufficient importance to compel us to solve it. In contrast to Melissa and the other chimp mothers, nearly all humans would be able to innovate a solution to the problem of psychopaths in their group if there were no law enforcement.

Necessity is often described as the mother of invention, but different people perceive their necessities differently, and it is the rare problem that demands a technical rather than a social solution. We can invent a trap to deal with murderous cannibals, but we can also enlist the aid of other members of our group who have been victimized by them to arrive at a novel, social solution for eliminating the threat. The capacity to innovate novel solutions appears to be universal in our species, but the proclivity to direct this capacity at technical rather than social solutions appears to be unusual. With just a little technical training and experience, nearly all healthy adult human minds are capable of creating technical solutions if the situation demands it. In the absence of such demands, technical innovation might be rare even if the abilities that underlie it are universal in our species. People who are interested in people (i.e., almost all of us) simply don't invent very many new things, regardless of their potential to do so.

Finally, although our inherent sociality might disrupt our tendency to invent new products, it plays a major role in transforming one person's invention into everyone else's solution. The human success story is not just one of innovation; it is also one of transmitting new inventions to others who use and improve them. Technical problem solving might be less frequent due to our incredibly strong social orientation, but sociality itself is critical for spreading technical innovation. The irony is that *Homo sapiens* rose to worldwide dominance due to our hypersociality, but it may be the relatively asocial ones among us whom we have to thank for the technical inventions that so differentiate our lives from those of all the other beasts on this planet.

7

Elephants and Baboons

THE EVOLUTION OF MORAL AND IMMORAL LEADERSHIP

Nelson Mandela is one of my heroes. In his early forties he was sentenced to hard labor for his efforts to overthrow the apartheid South African government, and he remained imprisoned until his late sixties. It's easy to imagine that he would emerge a broken or embittered man, and yet after he was released from prison, he became a force for democracy and racial reconciliation. Mandela easily won election as South Africa's first post-apartheid president, but after one term in office, he chose to step down rather than seek reelection (which he would have won just as easily). Mandela was as close to a saint as any politician I've ever seen, and certainly one of the great moral leaders of our time.

Robert Mugabe, of neighboring Zimbabwe, provides a sharp contrast to Mandela. Mugabe was also imprisoned for his role in trying to overthrow an apartheid government. Also like Mandela, Mugabe expanded social services and called for racial reconciliation when he was newly elected as prime minister. But the similarity ends there. When Mugabe faced internal competition from domestic opponents, instead of compromising with them as Mandela had done, he resorted to violence to consolidate and extend his rule.

Rather than stepping down after his first term, he orchestrated constitutional amendments that made his role more powerful and allowed him to run for repeated and rigged reelections. His policies devastated his country, leading to hyperinflation, unemployment, widespread disease, and food shortages.

Mandela and Mugabe shared a similar background and might have been motivated to act in similar ways—so why didn't they? The contrast between them raises the question of what prompts a person to become a moral or immoral leader. I couldn't have predicted the direction that either man would choose at the outset of his political career, and many people took far too long to condemn Mugabe because they couldn't believe what he had become. Although Mandela and Mugabe represent extreme examples, the psychological forces that created both men can be found in all of us.

The hypersociality we evolved to survive the vanishing rainforest suppressed our individualism but did not extinguish it. As a consequence, humans are a blend of self-serving and group-serving. On the self-serving side, evolution acts on individual organisms to maximize their inclusive fitness (the number of offspring they and their relatives successfully place in the next generation). As we saw in chapter 4, to attract a mate, people need to compete with others in their group, and thus any trait that helps people outcompete fellow group members will tend to be magnified in the population due to its impact on sexual selection.

This evolutionary pressure toward individualism and competition is met by the countervailing pressures placed on us by our role in cooperative and interdependent groups. We were successful on the savannah (and everywhere else), despite being small, slow, and weak, because we evolved the capacity to work incredibly well as a team. For this reason, teamwork and cooperation are highly valued in potential friends and mates, and our collective orientation benefits us individually by helping us gain coalition partners. People who are self-focused and uncooperative are less desirable, particularly as romantic partners, and hence sexual selection ensures that cooperation also remains common in the gene pool.

The tension between individualism and collectivism underlies human sociality and all our interactions with our fellow group members. This tension is exacerbated in the context of leadership, as the costs and benefits of group membership are magnified in our leaders. On the one hand, leaders depend upon, exemplify, and facilitate the group orientation that teamwork demands and that has made us so successful. Mandela's suffering on behalf of his people, and his subsequent willingness to stake his reputation on inclusiveness and reconciliation, are examples of this sort of group-serving orientation.

On the other hand, power brings benefits, and the benefits that accrue uniquely to group leaders are likely to nudge them to consider their personal and nepotistic desires ahead of their group's needs. Mugabe's willingness to starve his own people to consolidate his power and wealth is an example of this self-serving orientation. The resolution of this tension between group and self-orientation determines the morality of leadership. Moral leaders are those who act in their group's interest, and benefit from their decisions only to the degree that their group benefits. For reasons that will become apparent, I refer to such moral leaders as *elephants*. Immoral leaders are those who act out of self-interest and who benefit from their decisions at the expense of their group. I refer to such immoral leaders as *baboons*.

Elephants and Baboons

Adult male elephants are formidable beasts. With few if any natural predators, adult males choose to live either alone or in somewhat fluid groups of other males, and thus long-term elephant groups are composed exclusively of adult females and juvenile males and females. In principle, the strongest male elephant could impose himself as leader over the herd, but there is no incentive for him to do so. Their widespread plant food sources are not readily monopolized, and females announce their fertility to all males in their vicinity, so males must compete for sexual access whether they are in the group or reside kilometers away. Thus, elephant leaders are typically chosen from among the oldest females in the group, and this matriarch is relied on to coordinate group movements, migration, and responses to threats, such as lions. The leader's role in these situations is to call the other elephants to action and direct them toward threats or opportunities. She doesn't dash out in front to provide protection (when threatened by lions, all the adults position themselves in front to protect their young); nor does she suffer privations on behalf of her group. The leadership she provides is in the form of guidance.

Because leadership does not give her preferential access to food sources or mating opportunities, elephant leaders do not gain unique benefits from their position. Rather, good leadership in elephant groups provides equal and mutual benefit to all members of the group, and thus everyone in the group is motivated for the wisest elephant to lead the group. Group interests and individual interests are nicely aligned in this system, so elephant leadership is predominantly group-serving.

Baboon leaders sit on the other end of the continuum. As I discuss in chapter 1, savannah baboons are preyed upon by hyenas and large cats such as leopards and lions. For this reason, there is safety in numbers, and male and female savannah baboons live together in large groups. Although baboon troops rarely have leaders in the sense of an individual who provides guidance or protection to group members, males compete fiercely for dominance over one another. Females inherit their position in the dominance hierarchy from their mother, but males leave the troop of their birth to join another group, so they must rise through the ranks by dominating other members of the troop through continual threats or acts of aggression.

Because the alpha male typically provides no guidance to the troop, the goal of dominance over other baboons is entirely self-serving. The benefits available to other males (e.g., access to fertile females, preferred foods, and shady resting spots) are impossible to monopolize, but they are disproportionately taken by the alpha male. It is clearly not beneficial to other members of the baboon troop to be dominated in such a manner, but they are largely unable to prevent this outcome. Traits such as large physical size and aggression have evolved in baboons because they facilitate dominance of the troop, which in turn provides reproductive opportunities that permit these traits to flourish.

One takeaway from the contrasting leadership styles of elephants and baboons is that the distribution of resources matters. Similar to the predictors of territoriality discussed in chapter 3, baboons have a greater incentive to dominate their fellow group members than do elephants. With regard to food, the inability of herbivorous elephants to monopolize the abundant plants they require contrasts to the situation of omnivorous baboons. The relative scarcity of high-calorie foods (particularly meat) available to baboons imposes strong evolutionary pressures on dominance. Baboons compete fiercely for high-quality food, and control over food sharing allows baboon leaders to buy and maintain allies, which can be critical to maintaining alpha status.

Perhaps more important, differences in opportunities for reproduction also contribute to the type of leader that emerges. In part because elephant leaders are female, they experience no reproductive advantage as a consequence of their role. In contrast, approximately half of the reproductive success of male baboons is accounted for by their rank in the hierarchy, as dominant baboons have access to more mates than do subordinate ones. This difference imposes strong evolutionary pressure on male baboons to achieve a dominant position within the troop. Disgruntled troop mates can either challenge the alpha or try to join another group where they might be more successful. Either way, the troop provides protection and more eyes to detect predators, so nondominant males prefer group living over trying their chances alone.

Elephants and Baboons as Leadership Styles

Most human leaders are a blend of group-serving and self-serving, but people vary dramatically on where they fall on that continuum. Some people are mostly group-serving, like Mandela, and others are mostly self-serving, like Mugabe. An elephant-type leader maintains a group orientation under most circumstances, but baboon-type leaders are very sensitive to threats to their leadership, and they will shift between self- and group orientation depending on internal and external threats.

The clearest example of such situational shifting among human leaders can be found in the work of Jon Maner of Florida State University and his colleagues. In one set of experiments, Maner demonstrated that when baboon leaders (i.e., those high in dominance orientation) feel that their leadership position is threatened by subordinates, they limit information sharing with their group and exclude talented members, at a cost to group performance. Maner found that baboon leaders also rely on the tried-and-true strategy of divide and conquer when they feel their leadership position is threatened by talented team members. Such leaders try to prevent group members from bonding with one another.

The immorality of this leadership strategy is underscored by baboon leaders' efforts to exclude and isolate the most talented individuals. These actions place the leader's goals in direct opposition to the group's goals. More important, these immoral leadership behaviors disappear when baboon leaders are assured of their leadership position, again providing evidence that baboon leaders know how to enhance group performance but choose not to do so when their own stature is at risk. This sort of self-serving and group-harming behavior is a hallmark of baboon leaders across the globe, as countless despots have impoverished their people in their efforts to ensure that they remain at the top (of an ever-diminishing heap).

These immoral leadership behaviors disappear when baboon leaders find their group in competition with other groups. Under such circumstances, baboon leaders rally round to act in their group's interest, because group and leader goals (even baboon leader goals) are brought into alignment by the fact that failure in *inter*group competition can be very costly. Among our hunter-gatherer ancestors, a loss in intergroup conflict could lead to extermination of the entire group (or, more commonly, all the males in the group, with the females captured, enslaved, and eventually incorporated into the other group).

For this reason, we evolved a propensity to set aside our differences when competing with outside groups, to focus our energy on supporting our own group. Again, this sort of behavior can also be seen in despots around the globe, as many modern dictators began their careers as freedom fighters who often risked their lives and livelihood to remove the dictator or colonial leader who preceded them. But once such leaders are successful and intergroup competition is removed, baboon leaders place their own interests ahead of their group's and focus on consolidating and maintaining their power rather than serving their people.

Followers play a critical role in enabling baboon leaders under such conditions. As is discussed more thoroughly in chapter 8, deadly intergroup competition is a challenge that humans have faced for at least as long as we've been *Homo sapiens*, and it has shaped our psychology to accept and even prefer more dominant leaders during times of intergroup strife. A kind or indecisive person might be a good friend, but such an individual would be a hopeless leader in the face of a mortal enemy. Some elephant leaders rise to this challenge and change their leadership style to suit their situation, but some are simply too democratic for the demands of intergroup conflict. As a consequence, our leadership preferences change in the face of conflict. These psychological responses to within- and between-group competition represent adaptations to group living in a species that shows substantial cooperation within groups but often ruthless competition between them.

A Case Study: The Hadza and the Yanomamö

The Hadza are hunter-gatherers from Tanzania.* They live in relatively small and fluid groups of an average size of twenty to thirty people, who set up temporary camps for a few weeks or months at a time. The Hadza typically move to a new location when the local water hole dries up or they have exhausted nearby resources and must travel uncomfortable distances to gather foods. Due to their nomadic lifestyle, the Hadza own only as much as they can carry, so the sum total of their individual possessions is limited. Men usually own some jewelry and clothes, bows and arrows for hunting, and a few knives, axes, and such for setting up camp. Women also own jewelry and clothes, as well as digging sticks and cooking pots for gathering and cooking food.

The Hadza make all group decisions through discussion and have no explicit leaders. Older men and women are given a degree of deference but do not actually lead, and individuals who try to dominate others quickly find themselves shunned. Because individual Hadza can change groups at their own discretion, would-be leaders have little power to enforce their will over other people. The Hadza are relatively peaceful for a society with no law enforcement.

* This may have no relevance to their lifestyle, but the Hadza live in the cradle of humanity, on the same East African savannah where our ancestors were expelled from the trees.

They are also monogamous, and a couple is considered to have married when they set up house together. In his survey of childbirths, the superb ethnographer of the Hadza, Frank Marlowe of Florida State University, found that men average four to five children each, with a range from zero to sixteen.

The Yanomamö are hunter-horticulturalists who live in the Amazon basin of Venezuela and Brazil. Their villages also move on occasion, but because the Yanomamö do a fair bit of gardening (growing cassava and other foods), they move much less frequently than the Hadza. As a consequence, their domiciles are more permanent structures, and people tend to own more implements. Their villages also grow to a much more substantial size than those of the Hadza, sometimes containing more than three hundred individuals.

Most people prefer to live in smaller villages that are closer to the size of typical Hadza camps, and for the same reason that the Hadza prefer smaller groups: they lead to less bickering. The headmen in the smaller villages tend to lead by example rather than force, and life in small Yanomamö villages isn't that different from life in Hadza camps. Nevertheless, some leaders wield a great deal of power and prevent their groups from splitting apart, despite the wishes of their group members. Part of the motivation of such leaders is that many of the villages are in a near-constant state of war with one another, and larger villages have an advantage in these conflicts over smaller ones.

Many leaders among the Yanomamö are despotic and cruel. Such leaders hold group members in sway through the threat of personal violence, and typically rely on a network of male kin as their power base. Conflicts are commonly resolved through ritualized forms of violence. If there is a minor dispute, people arrange side-slapping and wrestling contests. For the side-slapping contests, the disputants line up in two different groups. Members of one group raise their arm in the air, and members of the other group slap them on the exposed side of their body as hard as they can.

The roles then reverse, and this plays out until everyone is sore and tired. At this point, the dispute is regarded as resolved, and people go about their business.

If there is a major dispute, the Yanomamö arrange the most extraordinary ritualized club fights. As in the side-slapping contests, people on one side of the dispute stand still while those on other side inflict damage, and then the roles are reversed. In this case, Person A stands with his arms by his side while Person B smashes a ten- to fifteen-foot pole down onto the top of A's head as hard as he possibly can. Person A may or may not crumple to the ground unconscious, but if he does, the next member of A's group stands in as a replacement and clubs Person B in a similar fashion.

This serial clubbing continues until all the members of both parties have been knocked unconscious or have decided that the grievance has been resolved. If a dispute arises between members of different villages that is too serious to resolve by clubbing, then a series of deadly tit-for-tat raids between villages ensues. Sometimes these raids end with a mutual feast as the groups resolve to put the past behind them, and sometimes they continue interminably.

Yanomamö are polygynous, often with little female choice, as powerful males have systems of trading their female kin with one another to obtain wives. Wives are frequently treated very poorly by such men, who physically abuse them at will, often in public and with no intercession from other group members. Violence appears to be an important route to obtaining leadership positions and wives, as men who have killed another man (termed *unokais*) have more wives than men who have not murdered someone. Commensurate with their greater number of wives, *unokais* also have more children than men who have not killed anyone.

Such differences in reproductive success provide an incredibly strong incentive toward violence and the baboon leadership that exists in some Yanomamö villages. Indeed, some Yanomamö leaders have huge numbers of children, and their remarkable ethnographer,

Napoleon Chagnon of the University of Michigan, documents one powerful leader as having 43 children and 229 grandchildren. Given the high child mortality rate among the Yanomamö, this is an extraordinary outcome.

Inequality and the Emergence of Baboon Leaders

Why are Hadza leaders (to the degree that such leaders even exist) primarily peaceful elephants while Yanomamö leaders are often despotic and violent baboons? No doubt many factors are involved, but the opportunity to monopolize resources appears to play an important role. Reminiscent of ever-grazing elephants, people in nomadic societies such as the Hadza encounter fewer opportunities for accumulating and controlling resources than those in more sedentary hunter-horticulturalist societies such as the Yanomamö. When resources can be controlled and used to leverage deference and compliance from others, it gives rise to the more competitive and self-serving aspects of our psychology. As we saw in chapter 3, the shift from hunter-gatherer societies to more settled horticulturalist practices was a critical point in the move away from egalitarianism and toward despotism in many human societies.

One important aspect of a horticultural lifestyle is the capacity to afford more than one wife. Yanomamö men who have killed another and are high in the dominance hierarchy often have many children through multiple wives, but this is not possible for immediate-return hunter-gatherers. Hadza men simply cannot hunt enough animals to support more than one wife, and even if they were epic hunters, they would be required to share their catch with the rest of the group, with limited opportunities to preferentially feed their family. Monogamy (and female choice) among the Hadza contributes to an elephant leadership style, whereas polygyny (and lack of female choice) among the Yanomamö contributes to a baboon leadership style. Indeed, the potential for polygyny plays a critical role in shaping the morality of leaders across a wide variety of societies.*

The enforced equality of Hadza society (with hunters required to share their kill and show humility, people owning few possessions, and reproductive success spread relatively evenly across society) means that there is little to be gained through despotic leadership. Not only do efforts to dominate meet stiff resistance from others (who have equal access to resources and are not dependent on would-be leaders), but there really are no additional resources to be garnered by such a strategy. Rather, wise counsel can help the group be more successful, and all individuals benefit to the degree that their group benefits.

Indeed, to the degree that individuals gain anything from offering their leadership during group decision making, it is through increased prestige by demonstrating wisdom and compassion. Thus, leadership in Hadza society can be obtained only momentarily and situationally, and the goals of would-be leaders must necessarily align with those of their group. The inherent selfishness that is part of human nature and the inherent group orientation that is also part of human nature are aligned by the egalitarian structure of Hadza society, and both push people toward elephant leadership.

In contrast, the Yanomamö marriage systems create the potential for substantial inequality. This inequality rewards men for their dominance and status seeking, as wives are obtained in part through violence against others, particularly other groups, and in

* Polygynous societies are often criticized for their poor treatment of women (as can be seen among the Yanomamö), but women in some polygynous societies, such as the Sukuma and Rangi of Northern Tanzania, are treated perfectly well. Women in such societies benefit from the greater wealth of their husbands, they and their children are typically well looked after, and hence they often choose to be second or third wives of wealthy men rather than first wives of poor ones. In contrast, as Robert Wright points out in *The Moral Animal*, poor men are always the losers in polygynous societies, as they spend their lives alone, unable to afford a wife.

part through alliances that can be more readily controlled by those in leadership positions. As a consequence, violent and despotic leaders have greater reproductive success than nonviolent individuals and people who are not in leadership positions. In this manner, the structure of Yanomamö society causes our inherent selfishness to oppose our inherent group orientation.

This motivational tension results in a leadership style that varies both with circumstance and with the personalities and proclivities of individual leaders. Particularly in smaller villages, group orientation is common, and leadership is often elephant-like. But because Yanomamö society has the potential for much greater inequality, a self-serving orientation is also rewarded, and the resultant leadership can be baboon-like, based on dominance over others rather than helpful guidance.

From Small-Scale Societies to Large-Scale Corporations

If we jump from forager societies to corporate America, we see many of the same inequality-based dynamics. Modern corporations are typically structured in a manner more similar to the Yanomamö than the Hadza, and as a result, they incentivize baboon leadership. For example, the average CEO salary in the largest American companies was $9 million in 2017. It's no surprise that competition for such a paycheck can be fierce. Nor is it a surprise that baboon leaders are particularly motivated by the enormous status, power, and financial reward of being a CEO.*

* I don't mean to suggest that all corporate CEOs are baboons or that really well-paid CEOs are necessarily baboons. Indeed, there are many notable examples of elephant leaders in the corporate world, some of whom are very well remunerated. But there are far too many corporate and political baboons (despite the fact that employees, shareholders, and voters would prefer it were otherwise), and the goal of this chapter is to discuss the psychology that underlies this unfortunate situation.

Although followers want to select elephants to lead them, baboons usually pretend to be elephants until they get the job.*

There are undoubtedly many factors that have produced greater CEO compensation over the last several decades. Perhaps most relevant to this discussion is the effect of the 1992 Securities and Exchange Commission ruling that decreed that U.S. companies must present CEO salaries in standardized tables rather than obfuscating them in the context of a long report with numerous details. The SEC was concerned about the fact that the ratio of CEO pay to that of the average worker in the company had risen from 20:1 in the mid-1960s to 100:1 in the early 1990s. The goal of this new format was to provide shareholders with clear information about pay disparities between CEOs and the average worker, and thereby shame CEOs into reduced compensation.

Unfortunately, the ruling had the opposite effect, and CEO salaries quickly skyrocketed to more than 200:1. As of 2017 they were at 130:1, but many CEOs on the list make four or five hundred times as much as their average employee (e.g., Comcast, T-Mobile, and Pepsi). Apparently, a more transparent salary presentation led CEOs to compete to earn more than their fellow CEOs. Thus, the assumption that CEOs would evaluate their compensation in comparison to their workers and be embarrassed by their excesses proved to be naïve. CEOs were more concerned about their relative ranking among one another. This finding suggests that CEOs were already predominantly baboons by the early 1990s, as in the competition to accrue the most resources, their orientation was not toward their group but, rather, was self-focused. This finding also

* Baboons also win followers by promising to bring them along on their quest for dominance. As Donald Trump repeated during his campaign, "We're going to win so much you'll be tired of winning." "America First" was an explicitly baboon leadership pitch: we will flex our military and economic muscle until we get what we want, never mind the cost to others.

identifies no clear ceiling in the desire of baboon leaders to benefit at the expense of their group.

Perhaps the reaction of CEOs to the SEC ruling should have come as no surprise, given the extraordinary willingness of dictators around the globe to impoverish their own people in their quest to enrich themselves. Daron Acemoglu and James Robinson's book *Why Nations Fail* is a tour de force through the world's failed economies and their causes, and along the way, it describes a depressingly large number of baboon leaders.

As but one example, consider Islam Karimov, the dictator of Uzbekistan until his death in 2016. When his disastrous economic policies ensured that farmers could no longer afford to maintain their cotton combines, he simply decreed that schoolchildren spend September and October harvesting cotton by hand. Children were required by law to harvest twenty to sixty kilos of cotton per day, depending on their age, and were paid about three cents a day. They were to bring their own food and find a place to sleep if they didn't live nearby. Through such baboon leadership, Karimov became a billionaire while the average Uzbek was forced to get by on a thousand dollars per year.

The takeover by baboons is not limited to the leadership structure of modern countries and corporations. Similar effects can be seen within societies, with inequality bringing out the baboon aspects of our psychology. When societies become increasingly unequal, people become increasingly desperate to be one of the haves rather than the have-nots. This desperation magnifies some of the more unpleasant aspects of human nature, one of which is self-enhancement, or the claim to be more than we really are.

The clearest example of this effect can be seen in a study by Steve Loughnan of the University of Melbourne and his colleagues, who found that self-enhancement increases in society as a function of income inequality. For example, Loughnan found that Japan (whose citizens are famously self-effacing for reasons long thought to be associated with their high levels of collectivism) anchors the low end of the self-enhancement continuum. Germans are much more individualistic than the Japanese, but they show similarly low levels of self-enhancement. As it turns out, Germany and Japan also happen to be countries with a high degree of economic equality.

In contrast, Peru and South Africa anchor the high end of the self-enhancement continuum and are also countries with high levels of inequality. These data suggest that income inequality can exacerbate our baboon psychology by leading people to claim to be much more than they really are. Because the consequences of failure in countries such as Peru and South Africa are so dire, people rely on every weapon in their arsenal to ensure they succeed. Inflated claims about one's worth emerge in such unequal societies as people strive to convince others of their abilities in an effort to be chosen for the few lucrative opportunities that exist.

One reason that self-enhancement features so prominently in baboon psychology is that it increases self-confidence, a quality that, for better and for worse, people strongly prefer in their leaders. In contrast to self-improvement, which leads to greater confidence via greater performance, self-enhancement leads to greater confidence via overconfidence. As we saw in chapter 5, people are largely incapable of distinguishing between overconfidence and high levels of well-calibrated confidence. As a consequence, people with unrealistically grand self-concepts have an advantage in leadership competitions. Such self-enhancing tendencies may help baboons secure leadership roles, but these tendencies do little to promote their effectiveness in such positions. Overconfident leaders often make poor decisions, ignore obvious flaws in their strategies, and continue with failing plans. Inflated self-views also lead baboons to over-reward themselves due to a trumped-up sense of their own abilities.

Not surprisingly, the results of baboon leadership are rarely good. When self-interest and dominance prevail, trust is hard to sustain, especially among the powerless, who are the first to be exploited. We have seen an erosion of trust in America in recent years that corresponds with sharp increases in inequality. Figure 7.1 shows that Americans had slightly higher levels of inequality and slightly lower levels of interpersonal trust than the people of Thailand when these data were collected in 2004. If we go back to the early 1970s, levels of trust and inequality in America would have placed the country about where Switzerland is in this figure. Not long ago, America was one of the most trusting countries; now it just sits in the middle of the pack.

Loss of trust matters because distrust leads people to disengage from their communities, reduce their commitment to their workplace, and reduce the degree to which they share important information with each other. In short, without trust, people focus on self-protection and become unwilling to make themselves vulner-

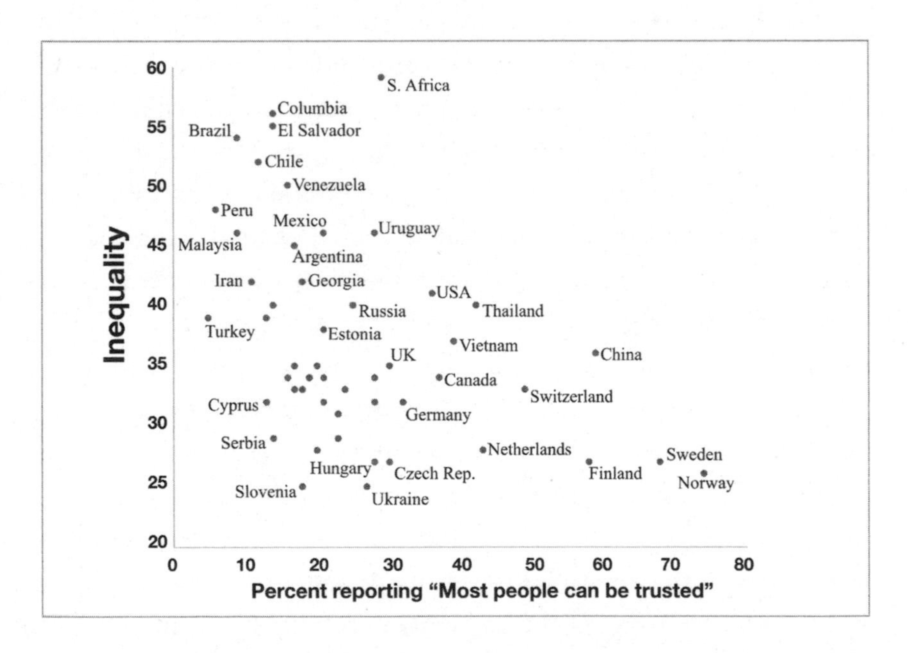

Figure 7.1. Worldwide trust and its relationship to income inequality. Higher numbers on the y-axis represent greater inequality, and higher numbers on the x-axis represent greater trust. (World Values Survey)

able. Elephant leadership becomes almost impossible under these circumstances, as group members can't align their interests.

Because vulnerability is an important ingredient in establishing and maintaining cooperative relationships, people who lack trust are limited in what they can accomplish together. To choose an ancestral example, the most critical factor in my decision to throw stones at an approaching lion (rather than run away) is my faith in my other group members to do the same. If I don't trust them, I have no choice but to run, which is a far inferior outcome. For this reason, the economies that have historically flourished are those in which trust in others is widely shared.

Can Anything Be Done to Enhance Moral Leadership?

This discussion of the evolutionary origins of moral and immoral leadership leads to a question about whether anything can be done to rein in baboonism and promote elephantism. Given the role of inequality, an obvious answer is to minimize it, particularly with regard to the rewards given to leaders versus other members of their group. Such changes aren't easy, of course, as baboon leaders fiercely resist efforts to enhance equality, but they can be achieved. For example, I suspect that CEO positions would still attract many qualified applicants if they did not pay an average annual salary of $9 million.

When efforts to reduce inequality fail, linking pay as tightly as possible to long-term performance outcomes creates a more elephant-like means of remuneration, as it aligns the interests of leaders to those of their group. Baboon leaders may not be as attracted to CEO positions that pay only twenty to fifty times the salary of the average employee (as was the norm prior to 1992), or that pay a salary that is closely aligned to group performance, but that is precisely the point. Elephant leaders are willing to take on the mantle of leadership even when the financial compensation for doing so is more modest or is entirely dependent on performance. The irony is that greater pay can lead to poorer leadership by attracting baboons to leadership positions and by turning leaders into baboons.

Beyond the unlikely solution of dramatically cutting the pay of CEOs and upper management, various other strategies can be adopted to minimize baboon leadership. Recall that Jon Maner found that conflict with other groups makes baboon leaders behave in a more moral fashion by putting their group's interests ahead of their own. This finding suggests that intergroup conflict can enhance leader morality. In modern corporations, this sort of conflict is the basis for the market economy, as corporate collusion is forbidden by antitrust legislation. But some corporations face greater threats from their competition than do others, and Maner's research suggests that the risks of baboon leadership will be higher in those corporations whose products and market position are more secure.

Unfortunately, competition with other groups is a double-edged sword. By raising the specter of such conflict, or by actually creating intergroup conflict where it need not exist, baboon leaders can generate greater group loyalty among their followers at the expense of group goals. Autocratic and dictatorial leaders throughout human history have used intergroup conflict and the creation of enemies of the state as a strategy to justify and bolster their baboon leadership. Of course, this strategy is itself an example of baboon leadership, as the creation of unnecessary conflict is often harmful to group goals. Thus, conflict between groups has the potential to enhance moral leadership while also serving as a cynical tool in the hands of immoral leaders.

Another solution can be found in the fact that ancestral groups involved relatively long-term membership in small bands. In large part, this was due to the relative sparseness of human populations and the low carrying capacity of the land until the advent of agriculture. With few (if any) alternatives, people typically remained embedded within the same small network of social groups for the entirety of their lives.

Knowledge of other group members in such a world was acquired either face-to-face or by word of mouth (i.e., gossip). These close-knit and more enduring social structures provided an efficient means of regulating the degree to which baboon leaders could negatively influence group outcomes. Enduring social relationships imposed strong costs on self-interested members looking to exploit the group for personal advantage. Enduring relationships aren't a guarantee against baboon leadership—one need only look to the Yanomamö to see that—but they do serve as a bulwark against would-be leaders who attain their leadership position by disguising their self-serving agenda.

In contrast, modern organizations, especially those in more individualistic Western cultural contexts, tend to be characterized by high levels of mobility. According to the U.S. Census, the average American moves nearly a dozen times, and the average time spent in a given job is only around four years. Indeed, it is now a relatively simple matter to move across entire countries or even around the world in search of a new job. In my own case, it didn't strike me as that big a deal to move halfway around the globe for a job in Australia. Technology allows me to stay in touch with friends and family, so why not live where I want?

The incredible mobility enabled by modern technology can cause people to be more concerned with their own goals than those of their organization, as they can leave their job behind if things don't work out. Although cultural differences play a large role in such decisions, our modern environment makes opting out of organizations and even entire communities easier than ever before. The result of this unprecedented mobility is an incentive structure that is well suited to the rise of baboons, who can exploit their organization and move on when the enterprise begins to suffer.

Given the limited mobility and relatively small pool of individuals in our ancestral groups, leaders emerged from within the rank and file of the groups they ultimately led. Thus, any individual's attempts to exert influence over the group would have been tempered by firsthand evidence of an aspiring leader's competence, knowledge, expertise, and commitment to the group. This internal approach to leadership selection is less frequently used in modern organizational contexts, where candidates (especially at the highest levels of the organization) are often recruited from competing organizations. External hires bring in new ideas and new energy, but hiring from the outside means that people who know the candidate are rarely involved in the leadership selection process. No doubt, this is part of the reason that externally appointed CEOs usually perform worse and have shorter tenures than those recruited internally.

Finally, modern organizations are at a disadvantage compared to ancestral groups, as the latter typically had a dynamic and distributed leadership structure, such that task expertise determined which individuals led in different domains. Although baboon types sometimes hold sway over hunter-horticulturalists, leadership among hunter-gatherers is typically malleable and context-dependent, often with little transfer from one domain to another. For instance, leadership in many small-scale societies includes war leaders, hunt leaders, medical leaders, song leaders, and peace leaders. Because task expertise is more readily judged than overall competence, ancestral groups with norms of task-based leadership would have been less susceptible to the false signals of baboons when choosing who should lead the group during specific activities.

In contrast, contemporary organizations have dominance hierarchies that cut across different activities, even when an organization includes divisions that rely on non-overlapping skill sets (e.g., engineering, accounting, sales). CEOs are the decision makers in charge of the entire organization regardless of the relevance of their expertise. As a result of these numerous shifts away from ancestral approaches to leadership, many American corporations and political entities are run by baboons, even though shareholders, employees, and voters are usually highly motivated to choose elephants.

This is not to say that our ancestors had some sort of wisdom that we have lost over the course of evolutionary time. There may well have been plenty of good ideas that have been lost, but the primary cause is the contrast between the environment in which we evolved and the environment in which we now find ourselves. This evolutionary mismatch is a source of many of our problems, and as we will see in the next chapter, it greatly increases the risk of unnecessary conflicts that could conceivably annihilate us all.

8

Tribes and Tribulations

EVOLUTIONARY PSYCHOLOGY AND WORLD PEACE

We are incredibly fortunate to live in a time of increasing peace and diminishing interpersonal violence. It may not feel like it if you keep up with the news, but over the last thousand years, hundred years, and even twenty-five years, violence has decreased across the industrialized world. There are many reasons for this fortunate state of affairs, as the establishment of democracy, strong institutions of government, international trade and tourism, and numerous other factors have contributed to our greater understanding of one another and a more peaceful world.

Decreased interpersonal (as opposed to interstate) violence is largely a product of two factors. First, the presence of strong and relatively impartial law enforcement has undermined the three main sources of violence identified by the great political philosopher Thomas Hobbes. In the presence of a strong and impartial state, I'm no longer motivated to attack you to take what's yours, as I know that the state will punish me. It's also no longer necessary to attack you to prevent you from taking what's mine, because, again, I know the state will protect me. And finally, I no longer need to retaliate if you attack me or my family, as I know the state will punish you.

Second, and more speculatively, as the industrialized world has become a safer place, those of us who are fortunate to live in it have become more sensitized to violence. Increases in safety happen slowly, so we tend not to notice them, but over a long enough time span, the effects are dramatic. The biggest difference, of course, is that between hunter-gatherers and people who have access to modern medicine. Consider the survival curves depicted in Figure 8.1 documented by Brian Wood of Yale University and his colleagues, showing chimpanzees in different parts of Africa and humans from different foraging societies.

As is evident in the left panel, nearly half the chimps born across these various groups die before the age of five (which is still childhood for a chimp). Rather remarkably, the survival rates aren't that much better for humans: 20 to 40 percent of human children also die by age five (right panel). If we consider twenty years old to be the beginning of adulthood in humans, we see that 25 to 50 percent of humans across these societies never reach adulthood. High levels of violence in forager societies contribute to the high mortality rate, but the high mortality rate undoubtedly contributes to the violence as well. When people become

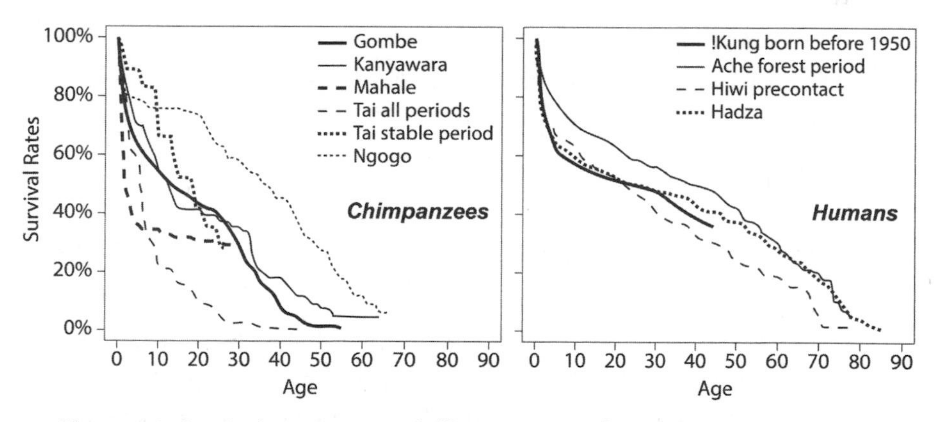

Figure 8.1. Survival rates among different groups of chimpanzees and human foragers. (Adapted from Wood et al., 2017, by Gwendolyn David)

inured to suffering, the harms they cause each other seem less significant.

It is no surprise that our lives have become much safer than when we were hunter-gatherers, but changes in safety are also substantial in modern environments. Figure 8.2 depicts changes in mortality rates during childbirth for mothers and during the first five years of life for children in the United States across the twentieth century. As is evident in these data, childbirth and childhood have become dramatically safer. In the early 1900s,

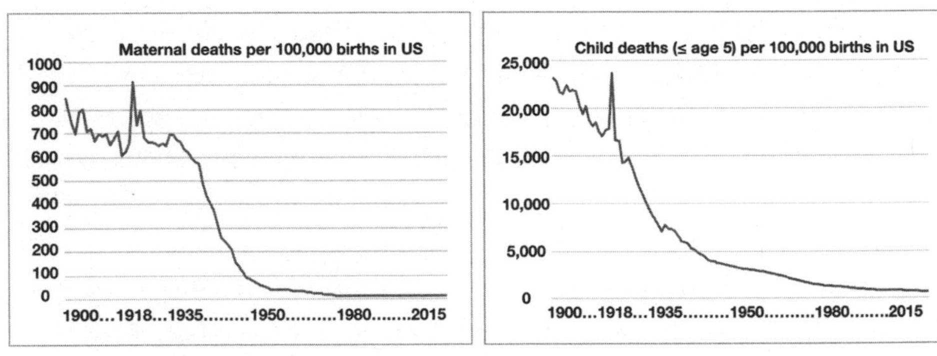

Figure 8.2. Maternal (*left*) and child (*right*) mortality rates in the United States. (Our World in Data)

nearly 1 percent of mothers died in each delivery. Even more remarkably, almost 25 percent of children died before the age of five (a rate equivalent to those of some of the hunter-gatherer groups depicted in Figure 8.1). In such a world, everyone knew someone who died from complications in delivery and who lost children before their fifth birthday.

Numerous other advances in health mirror these positive trends. For example, at the start of the twentieth century, 120 out of every 100,000 people in the United Kingdom died of food poisoning each year. To put this number in perspective, it was more dangerous to go out for fish and chips in Oxford in 1890 than it was to hang out in downtown Detroit a hundred years later. Similar statistics can be found for numerous other threats to health and longevity, and many didn't abate until very recently.

Perhaps these increases in health are unsurprising, as modern medicine has been such a success story. But other data show that changes in safety are not just a function of medical advances—people's behavior toward each other has also changed. For example, we can see big drops in the homicide rate across America's five largest cities (New York, Los Angeles, Chicago, Houston, and Philadelphia) over the last thirty years. In the late 1980s to early '90s, these cities all suffered an average of about 30 homicides per 100,000 citizens per year. By 2015, New York and LA were at less than a third of their prior homicide rate, Houston and Philadelphia had cut their earlier rate by half, and only Chicago's recent rise in crime resulted in murder rates that are anywhere near what we saw in the '90s. These data show that living in big American cities has become *much* safer in just the last thirty years.

Murderers may not be like everyone else, so you could reasonably argue that these data don't speak to widespread behavioral change. For evidence of widespread change, drunk driving is one of the best examples. As recently as the late '70s, drunk driving was not only commonplace but perceived as an unavoidable fact of life. Perhaps the most remarkable thing was that people treated it like a joke. When someone had too much to drink at a party, their friends would go so far as to help them insert the key in the ignition and then laughingly watch them lurch off down the road. In *Fear and Loathing in Las Vegas*, Hunter S. Thompson describes how a police officer let him drive home to sleep it off when he was clearly intoxicated behind the wheel. That event may or may not have actually happened, but it seemed perfectly plausible at the time.

Everything changed with the advent of Mothers Against Drunk Driving in 1980, which can be credited with drawing awareness to the (admittedly obvious) fact that drunk drivers hurt people other than themselves. Since 1982, when DUI statistics were first recorded (a remarkable fact in itself), drunk-driving deaths in the United States have dropped from more than twenty thousand per year to under ten thousand. This improvement is all the more noteworthy when you consider that Americans have doubled the number of miles they drive since 1982, suggesting that four times as many drunk drivers were on the road in 1980 as there are today.

Along with these massive changes in safety came massive changes in attitudes. My favorite example of how much attitudes have changed took place when I was on my way home from kindergarten in 1969. My friend Andy's father, a police officer, happened to be driving our carpool that day. There were four or five of us sitting in the backseat of his car, and none of us had seat belts on because cars were rarely equipped with seat belts in those days. As we rounded a corner a few blocks from my house, all the children slid across the bench seat and pressed against me. I tried to push back, but they were too heavy and I was pressed against the door. The next thing I knew, the door flew open and I was bouncing across the road and into a ditch. Andy's dad stopped the car to retrieve me—no surprise there; he could hardly have arrived at my house empty-handed—but what might seem surprising was my mother's reaction when he brought me home.

Take a moment to consider how you would react if a friend were driving your five-year-old home from school and showed up with the embarrassing admission that your child had fallen out of the moving vehicle and rolled across the street. Even if your friend quickly added the good news that none of the subsequent vehicles ran your child over, you'd probably be horrified and nearly speechless.

In contrast, my mother looked over my bumps, cuts, and scrapes and told Andy's dad that I seemed fine and not to worry about it; these things happen. Lest you think my mother was callous and uncaring, when my friend's brother fell out of his family car under similar circumstances, his mother berated him for not holding on more tightly. This is the world we lived in not so long ago. Drunk drivers were an amusing nuisance, no one wore seat belts (or even imagined a world of children in car seats), and everyone was fine with that.

In such a context, homicide or assault would have been a tragedy for the victim's family and friends, but to everyone else such crimes were an almost unnoticeable blip against a background of so much death and mayhem. As time passes and we become accustomed to food that doesn't kill us; seat belts, shatterproof windshields, and airbags that protect us; and dogs that don't bite us (because they no longer roam neighborhoods freely), each individual incident begins to stand out in sharper relief.

Opponents of violent TV and violent video games argue that these forms of pseudo-violence inure us to the real thing, but I think the negative effects of violent entertainment are dwarfed by the increasingly safe world in which we reside. There is no doubt that the internet allows us to see real carnage whenever and wherever it occurs in the world, and I'm sure that this contributes to our sense that the world is more dangerous than it was when we were younger. But our everyday experience is one of safety, and I suspect that our softened sensibilities have played a self-perpetuating role in making people less likely to commit murder, rape, and assault over the last few decades.

The increasing safety and decreasing violence that we see *within* industrialized countries have been mirrored by decreasing levels of conflict *between* countries as well. Nevertheless, there are constant threats to global security: conflicts in Europe, the Middle East, and Africa frequently escalate; terrorist groups such as Islamic State seek to inflame religious tensions; many countries are nostalgic for a time when they wielded a larger influence on the world stage, and make territorial claims that are inconsistent with the current international order; and, perhaps most surprising of all, xenophobic politicians show enduring (and at the time of this writing, increasing) popularity even in wealthy and stable countries. For these reasons, international peace and security cannot be taken for granted. The remainder of this chapter outlines how our evolved psychology contributes to this tenuous state of peace on the planet.

Humans Evolved to Cooperate within but Not between Groups

Chapter 1 discusses how we survived and thrived on the savannah by learning to be much more cooperative than our chimpanzee cousins. In principle, cooperation has the potential to make us more peaceful, but it turns out that our cooperation is highly selective; we evolved to cooperate with our fellow group members, but not with members of other groups. The reason for this state of affairs is that in our ancestral past, other groups were a toss-up: sometimes an opportunity and sometimes a threat. When encounters with other groups proved to be friendly, our ancestors increased their mating opportunities (and decreased inbreeding) by finding partners in other groups. But encounters often proved to be unfriendly, as sometimes members of other groups wanted what we had and chose to take it by force.

Stone throwing dramatically increased our safety on the savannah, but it also opened up an entirely new threat: the power to kill at a distance marked the beginning of highly effective warfare. Once we gained the ability to engage in collective stoning, we also gained the ability to engage in pitched battles. When *Homo erectus* doubled down on that capacity and developed planning and division of labor, strategizing and military campaigns became possible. As a consequence, the selective cooperation only with members of our own group was a prudent choice for our ancestors, as other groups rapidly became our most dangerous competitors. Once our ancestors moved to the top of the food chain, the risks posed by other animals rapidly diminished, and humans soon became their own greatest threat.

As I discuss in chapter 7, people in small-scale societies are often in a state of conflict with other groups. Such conflicts typically manifest themselves in skirmishes and raids rather than the pitched battles we associate with warfare between industrialized nations. But skirmishes and raids are also deadly, particularly over the long term, and members of successful war parties benefit from their violence through the theft of transportable wealth such as cattle and the opportunity to gain additional wives in the form of captured women. Intergroup conflict has thus been sustained in part because it can lead to greater reproductive success for members of successful war parties.

In addition to the competition for resources posed by members of other groups, our ancestors also faced the risk that other groups had been exposed to different pathogens, and thus could potentially give them new diseases. In a premedical world, the threat of disease was much greater than it is today, so we developed psychological adaptations to disease threats that are collectively characterized as the behavioral immune system. Our biological immune system deals with pathogens once we have ingested them (e.g., by vomiting them out or finding and destroying them), but our behavioral immune system evolved to prevent us from ingesting pathogens in the first place. For example, feces, open sores, infections, and vomit are rife with pathogens, so we evolved to find the sight and smell of them disgusting and to avoid them. People who found other people's open sores interesting or attractive were far less likely to survive and have children of their own, and thus evolution ensured that our disgust protected us.

Pathogens vary a lot in their virulence, and our psychology is fine-tuned to avoid those germs that are the greatest threat to us. By way of example, consider the following thought experiment. Imagine that you're on an airplane and standing in line to use the toilet. When your turn arrives, you notice that the person exiting the bathroom has a rather sheepish expression. You go inside and raise the toilet lid, and the cause of his expression immediately becomes obvious: evidence of his gastrointestinal distress is everywhere.

Here's the question: would this situation be more disgusting if the guy with tummy trouble were your brother or the random stranger seated in 21C? Most people find the scenario more revolting if the culprit is from 21C. And there's good reason for this: a stranger's feces are much more likely to make you sick than your brother's feces. You have far greater prior exposure to your brother's germs than to a stranger's, so you are more likely to have developed immunities to them.

A version of this thought experiment has even been run with actual poo, and the results are as predicted. The intrepid Trevor Case of Macquarie University and his colleagues conducted a study in which they asked mothers of young babies to come into the lab with their child's poopy diapers. The researchers separated the poo into containers and then asked the mothers to sniff them. The mothers were incapable of detecting which poo came from which baby, but they found the smell of the other babies' poo more disgusting than that of their own baby. Even though mothers were unable to identify their baby's poo, at an unconscious level their behavioral immune system pushed them away from the feces with a higher level of unfamiliar pathogens.

Experiments such as these point to the exquisite sensitivity of the behavioral immune system, and our evolved capacity to avoid germs that are most likely to make us sick. We see additional evidence for these processes in the geographic distribution of languages, religions, and ethnocentrism. As we move from the poles to the equator, the number of languages and religions per region increases, and people become more xenophobic. These effects may seem to be unrelated, but all three processes serve to keep groups apart. When you don't speak the same language, when you don't share a religion, and when you tend to dislike members of other groups, you're much less likely to intermingle with them.

Why would languages and religions proliferate around the equator, and why is their frequency also related to ethnocentrism? The answer to these questions lies in the fact that pathogen density is much higher in the tropics than it is in temperate and cold climates. When you live in Sweden, chances are good that any group within five hundred miles has been exposed to the same few pathogens. In contrast, when you live in the Congo, the group on the other side of the valley may well have been exposed to a pathogen with which you've had no prior contact.

For this reason, humans in the tropics learned that when they interacted with other groups they tended to get sick, so they would have stopped doing it. In a pre-scientific world, it was logical to blame their neighbors for their illness (which was partially true, at any rate), and therefore to dislike them. Dislike and fear kept neighbors apart, and once you don't interact with others anymore, your languages and religions naturally diverge as well. All these processes are self-perpetuating and serve to further group separation.

The effects of pathogens on attitudes, behaviors, and beliefs that keep groups apart also tend to be stronger when the pathogens rely on human-to-human transmission (e.g., hepatitis) than when they are transmitted from animals to humans (e.g., malaria). When other people engage in cultural practices that differ from our own, their behaviors not only call into question our cultural or religious practices, but also have the potential to lead to different routes of disease transmission. Different ways of cooking, different rites of passage, and different mating systems can lead to new forms of pathogen exposure. Our behavioral immune system evolved to ensure that such foreign practices seemed not just different to us, but also *wrong*.

Although our attitudes evolved to help protect us from illness, the attitudes that make up the behavioral immune system aren't directed at the germs themselves. Rather, these attitudes simply caused us to avoid dangerous pathogens, the sources of which we didn't understand when the attitudes were evolving. For example, by regarding behaviors different from our own as wrong or immoral, we tend to stay away from people who engage in those behaviors, which protects us from them.

Because humans tend to moralize behaviors that play an important role in procreation and disease transmission (which are often the same behaviors), merely different ways of doing things rapidly become immoral ways of doing things if they have the potential to serve as a disease vector. It is thus likely that pathogen threats are an underlying source of what is known as symbolic prejudice, or the animosity toward groups with different practices and beliefs from our own. Once other groups are perceived as immoral, people are much more likely to avoid each other, and when they do come in contact, conflict is much more likely to ensue.

For all these reasons, we evolved to cooperate with members of our own group, but not with members of other groups. This tribalism is often perceived as inconsistent with our cooperative nature, but our evolutionary history reveals that they are in fact two sides of the same coin. Our tribalism is actually cause and consequence of our cooperative nature, as our capacity to care for members of our own group evolved to make us more effective killers.

Evidence for our selective, within-group cooperation can be found in numerous places, but perhaps the clearest distinction is the one between humans and chimpanzees. To obtain an indicator of our historical rates of conflict, Richard Wrangham of Harvard University examined levels of physical conflict among human hunter-gatherer populations. Because they have no recourse to formal laws or police, hunter-gatherers provide the best indication of what our species level of violence was like before the advent of modern government. When Wrangham compared the level of within-group conflict among hunter-gatherers to that of chimpanzees, he found that chimpanzees were 150 to 550 times more likely than humans to resort to physical aggression within their groups. In contrast, when he examined acts of between-group aggression and violence, he found that rates of violence are basically the same across human foragers and chimpanzees. Our move to the savannah made us much nicer to each other, but this effect doesn't translate across group boundaries.

Despite these evolutionary changes, conflict and competition remain rife within groups, as people often fight over resources or disagree about how to solve communal problems. Sexual selection creates the most important challenge to within-group cooperation, as everyone is motivated to increase their status relative to other members of their group. These are the reasons hunter-gatherer groups are often overcome with bickering when they become larger than twenty or thirty individuals. Nonetheless, the threat posed by other groups was a substantial countervailing force to conflict within groups. The predilection to cooperate with each other when threatened by other groups was critical to our survival, as between-group competition was an existential threat whereas within-group competition was merely a status threat.

These competing aspects of our evolved psychology continue to manifest themselves in important ways in national and international relations. Perhaps most notably, cooperation within groups can dissolve over time in the absence of an external threat. The extreme partisanship and toxic political climate in the United States have been explained in many ways, but from an evolutionary perspective, part of the underlying cause is the demise of the Soviet Union. Conflict between political parties was once set aside at the nation's border because internal conflicts were less important than maintaining unity in the face of a powerful enemy. The fall of the Soviet Union marked the end of the only real existential threat to the United States, with the consequence that internal conflicts are now less likely to be held in check. Without the countervailing threat of a sufficiently powerful external enemy, the political parties increasingly see their greatest obstacles not in the actions of other countries, but in the competing goals and preferences of their domestic rivals.

The impact of external threats can easily be seen in Figure 8.3, which depicts how rapidly Americans came together in their support of President Bush after the attacks of September 11, 2001. Figure 8.3 also shows how that surge in support slowly dissipated as the threat came to be seen as less imminent, and then spiked again to a lesser degree when the United States invaded Iraq.

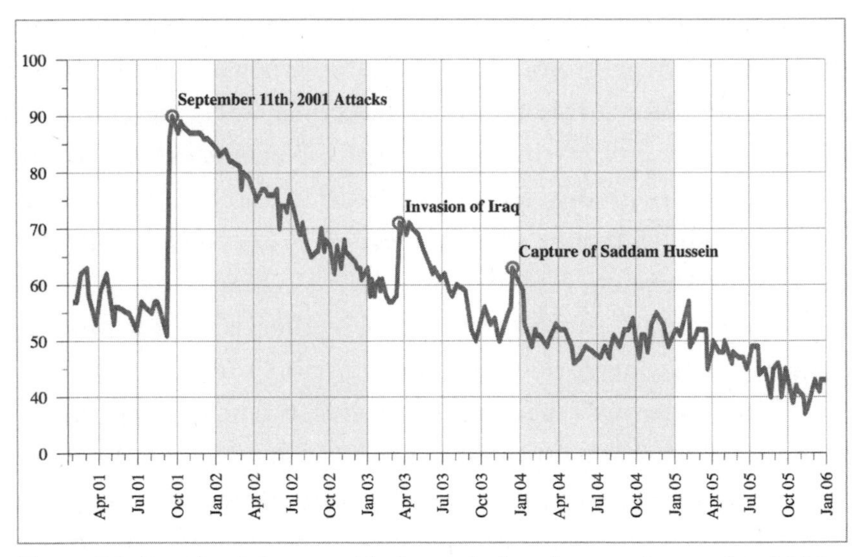

Figure 8.3. Presidential approval before and after the terrorist attacks of 9/11 and the invasion of Iraq. (Wikimedia Commons)

As discussed in chapter 7, our leaders often exploit this aspect of our evolved psychology, highlighting the potential threat posed by other groups in an effort to turn their followers' attention away from the group's internal problems or their own poor leadership. This strategy enhances loyalty and cooperation within the group, thereby strengthening the leader's position, but these gains come at an ongoing cost of disrupted relationships with other groups (resulting in lost commerce and increased conflict). Thus, our cooperative nature can actually lead to greater conflict between groups, particularly when group leaders value their privileged position in the group more than they value the group's goals.

Despite the substantial obstacles that these aspects of our evolved psychology pose to intergroup cooperation, even mutually suspicious groups can and do cooperate with each other to achieve overarching goals such as trade and security. Current and historical small-scale societies show numerous instances of intergroup cooperation, often in coalitions against other, more powerful groups, but also in service of intermarriage and trade. As a consequence, intergroup attitudes are most accurately described as an automatic bias in favor of our own group coupled with an ambivalence toward other groups.

This ambivalence manifests itself as a readiness to dislike or like other groups depending on whether they are currently perceived as a threat or an opportunity. The default presence of ambivalence rather than negativity toward other groups is critical, however, as it allows us to notice opportunities when they emerge and form mutually beneficial cooperative ventures across group boundaries. Treaties and alliances are thus most effective when all parties share a common goal so that the intergroup threat is defused and cheating is perceived as impossible or of no potential benefit (more on this point next).

Relativity Disrupts Intergroup Relations

As I discuss in chapter 4, our sense of fairness is driven by the comparison between our outcomes and those of others in our social network. This relativity emerges from the fact that position in the status hierarchy is a critical determinant of one's attractiveness as a mate. Status differences *between* groups are also important determinants of individual outcomes. By the time *Homo sapiens* began colonizing the distant reaches of the globe, we were in frequent conflict with other groups over resources. Indeed, much of human history has been a story of stronger groups pushing weaker groups away from preferred hunting grounds, watering holes, and fishing sites. We marvel at the audacity of our ancestors who struck off into the unknown, but a fair portion of human expansion and exploration reflects desperation rather than bravery. As often as not, exploration meant that our ancestors were escaping, or being forced to abandon preferred areas by their stronger or more aggressive neighbors.

Many scholars hold a romantic view of our past and like to think that human violence is a product of our modern life and of the disconnection and anomie created by urban living. But as Steven Pinker has shown so clearly in *The Better Angels of Our Nature*, this view is false. If we were so nice to one another prior to the advent of modern living, why do ancient human remains show so many projectile wounds? For that matter, why are injuries to the left side of the body so common? Rocks don't fall disproportionately on the left side of our heads, or our left arms, or the left half of our rib cages. But weapons wielded by right-handed opponents do, so it's no wonder many of our ancestors met their demise from wounds to the left side of their bodies.

Evidence for hostility between groups can also be found in the ancient cliff dwellings that dot the American Southwest and various other places around the world. When you visit these sites you wonder why anyone would choose to have babies and raise a family on the side of a cliff when they could have a much less precarious life in the valley below. The answer, of course, is that people chose to live on cliffs in Colorado and New Mexico because the cliffs were safer than the valley below.

Valleys provide little risk of falling, but when their inhabitants are nasty, they can be very dangerous places. Cliff-side living arrangements provide protection against attack, but at an ongoing cost of life and limb, particularly when people accessed their cliff dwellings via rickety ladders at the best of times. These dangerous living arrangements provide evidence of escape-driven expansion and show the cost to one group of people when another group becomes comparatively more powerful. For this reason, relative fairness can matter as much between groups as within them.

Due to the importance of relative fairness, humans have evolved a hypersensitivity to the possibility of being cheated. This sensitivity can be seen in various aspects of human psychology, one of which is an enhanced capacity to solve problems when they are phrased in terms of detecting cheaters compared to when they are phrased in terms of general rules of logic. Consider the following problem adapted from experiments by Leda Cosmides of UC Santa Barbara and her colleagues. In Figure 8.4 a series of cards indicate what people are either eating or drinking for breakfast. One side of the card shows what the person is eating, and the other side shows what he or she is drinking. Your task is to turn over as few cards as possible to test the accuracy of the rule "Everyone who eats cereal also drinks orange juice."

Figure 8.4. Rule testing: "Everyone who eats cereal also drinks orange juice."

As you can see, Jordy is eating cereal, and Maya is having pancakes, but we can't see what they're drinking. Sam is drinking orange juice, and Sophia is having coffee, but we can't see what they're eating. If you wanted to test the rule that everyone who eats cereal also drinks orange juice (by turning over as few cards as possible), whom would you check? Would you look to see what Jordy and/or

Maya are drinking? Or might you check what Sam and/or Sophia are eating? According to Cosmides and her colleagues, this should be a relatively difficult problem to solve, as it relies on formal logic. Take a moment to decide for yourself.

Before we discuss the answer to this question, let's consider an alternative version of the problem that Cosmides also gave her participants. For this alternative problem, your task is to test whether people are breaking the rule "People aren't allowed to drink cassava juice unless they have a tattoo."

Here we can see that Jordy is drinking cassava juice, and Maya is having orange juice, but we can't see whether they have tattoos. Sam has no tattoo, and Sophia has a tattoo, but we can't see what either of them is drinking. If you wanted to test whether any of them was breaking the rule that you must have a tattoo if you want to drink cassava juice (again, turning over as few cards as possible), whom would you check? According to Cosmides, this should be a comparatively easy problem to answer. Take a moment and see what you think.

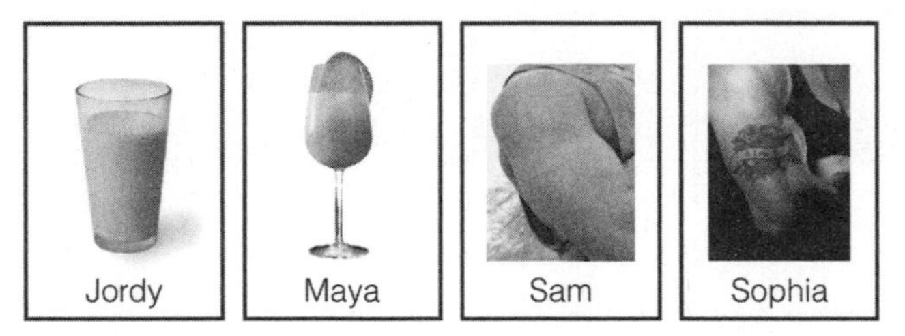

Figure 8.5. Rule testing: "People aren't allowed to drink cassava juice unless they have a tattoo."

Returning to the first problem, if you're like most participants in Cosmides' experiments, you probably checked Jordy to be sure he *is* drinking orange juice, but you probably failed to check Sophia to make sure she *isn't* eating cereal. Given that she's drinking coffee, if she's eating cereal, she's violating the rule that everyone who eats cereal also drinks orange juice. Consequently, checking both these people is critical to detecting if the rule has been violated. It doesn't matter what Maya is drinking, as no one said people who eat pancakes can't have orange juice, and it doesn't matter what Sam is eating—again, for the same reason. Even when people get this problem right, it takes a fair bit of concentration to work out whom to check and who can be ignored.

In contrast, most people find the second problem about cassava juice and tattoos pretty straightforward, because it taps into our evolved propensity to be on the lookout for cheaters. People typically test Jordy to be sure he has a tattoo, and they test Sam to be sure he's not sneaking cassava juice when he's not allowed. They don't bother testing Maya, because the rule isn't relevant to her, and they don't care what the tattooed Sophia is drinking because she's allowed cassava juice if she wants it, but of course is not required to drink it.

The very same problem that people struggle to solve when it's framed in terms of logical rules suddenly becomes easy to solve when it's framed in terms of following rules. We aren't very good at formal logic, but evolution has ensured that we are very good at detecting cheaters. This experiment suggests that we are chronically alert to the possibility of cheating, which enables our cognitive machinery to function more effectively in such contexts. And perhaps most important for intergroup relations, this increased cognitive sensitivity to cheating disappears when people are in situations in which they believe cheating is impossible.

So how do concerns about relative fairness and cheating impact efforts to attain peace and global security? The unfortunate upshot of these concerns is that when groups, societies, or nations attempt to negotiate peace treaties or even trade agreements, they are hamstrung by the desire of both sides to ensure that the other side is not getting a better deal. Because fairness concerns are *relative*, it is not sufficient for an agreement to benefit both sides compared to the status quo. Rather, agreements must not be seen to benefit one side more than the other. Concerns with relative status ensure that people will reject a treaty, even one that is better than their current arrangement, if it is perceived as bringing greater benefits to the other party.

This concern about relative outcomes is compounded by hypersensitivity to the possibility that the other side might be cheating and not meeting its obligations under the agreement. Because people are so sensitive to cheating, they tend to overreact when they see any hints of it, and quickly stop meeting their own obligations so that they do not provide others with a relative advantage. For example, if two countries agree to a moratorium on nuclear testing, but they suspect each other of cheating, they are both likely to start testing in secret to avoid falling behind in a nuclear arms race. As a result, any system of agreements can rapidly collapse if there is insufficient ability to detect or insufficient authority to punish cheaters.

International rules can be created to punish cheaters, but sovereign nations are typically loath to agree to supranational treaties that include sanctions that could be directed at them. To choose a recent example, China ratified the United Nations Convention on the Law of the Sea in 1996 but refused to accept the 2016 ruling under that law by the international tribunal in The Hague, denying the validity of its claims in the South China Sea. The Chinese Foreign Ministry released a statement noting that the decision "has no binding force." China is far from unique in this regard, as a similar reluctance has been shown by the U.S. government to allow international bodies to have jurisdiction over its actions. There are exceptions to this rule, as exist in many trade and armaments agreements, but such treaties often follow rather than precede the establishment of trust, or they emerge in domains in which rivals' interests align.

For these reasons, technical solutions that make cheating impossible can enable agreements to succeed that would otherwise fail through mistrust. For example, countries might easily lie about their nuclear-enrichment activities, leaving other countries to wonder if perhaps they should secretly engage in their own enrichment. In response to this situation, physicists have shown that commercial imaging satellites can detect plutonium production through changes in the atmosphere, thereby providing clear evidence if a country is cheating on its treaty obligations. Such evidence can forestall an arms race that nobody wants but everyone feels compelled to join lest they fall behind.

Humans Evolved to Be Self-Deceptive Hypocrites

Predator-prey interactions must eventually end in death: one becomes dinner or the other starves. In contrast, competition between members of the same species usually involves elaborate signaling that is intended to clarify which individual is stronger *before* conflict escalates to the point of injury. It is in neither the winning nor the losing party's interest to fight over resources, as even winners get hurt in fights. Consequently, it is in both parties' interests to determine who would win if the fight were to take place, with this determination followed immediately by deference or flight by the party destined to lose. Only when competitors appear to be equally matched does a conflict over resources escalate into actual physical combat.

For this reason, competition among members of the same species rarely relies on actual force, as the threat of force is a deterrent to conflict. Human conflicts are guided by the same principles as those of other animals, but the fact that warfare is so common testifies to the frequency with which the two sides are unable to agree on who would win a fight.

Because both parties typically suffer costs when a competition escalates to violence, contests between members of the same species are typically a blend of truth and exaggeration by each party intended to convince the other party to back down. Exaggeration would disappear if there were no cost to testing one's abilities against those of one's opponent. If we're competing over the last slice of cake and I think I might be stronger than you, I'll just punch you and find out. But there is a notable cost to this test, as you are likely to punch me back—a bummer under the best of circumstances, but particularly so if you're stronger than I am. It is this guaranteed cost of competition that allows deceptive individuals to exaggerate their strengths and downplay their weaknesses without necessarily getting caught. This type of exaggeration can be seen throughout the animal kingdom, such as when moose or hyenas raise the hackles on their back to appear larger, or when crabs grow unnecessarily large claw shells that they do not fill with muscle.

As we saw in chapter 5, self-deceptive overconfidence is also ubiquitous among humans, who typically believe they are better, stronger, faster, and more attractive than they actually are. This overconfidence plays a critical role in conflict escalation, as it causes the eventual losers of a conflict to believe that they will emerge victorious. For example, both sides in the American Civil War believed they would win in just a few months and with limited casualties. These self-deceptive beliefs led politicians on both sides to orchestrate the deaths of more than six hundred thousand of their fellow Americans. In combination with an evolved lack of cooperativeness with other groups, the belief in one's own likely victory is a driving force toward conflict.

Self-deception plays an important role in the failure of the losing side to predict its eventual loss, but the advent of nuclear weapons seems to have created a new reality in which people appreciate the deterrence value of such extraordinary force in the absence of using it. The primary advantage of nuclear arms appears to lie in the realization that even the winning side will suffer intolerable losses, and thus both sides can continue to predict their eventual victory but choose not to escalate.

Self-deception helps people believe not only that they are stronger or smarter than they actually are but also that their own actions are more moral than the apparently identical behaviors of others. For example, if I take a second helping of cake and you end up with none, I can assure myself that my apparent selfishness was just an oversight, as I didn't notice that you hadn't eaten dessert yet.* If you do the same thing to me, I'll be appalled at your poor manners and gluttony.

Because I'm inclined to recast my own motives in a positive light, even if you and I go through life engaging in the exact same behaviors, I will regard my actions as more moral than yours. This hypocrisy is fundamental to human nature, and like self-deception more generally, it evolved to help us persuade others that even when our actions aren't exemplary, our hearts are in the right place. This hypocrisy extends to other members of our group, as we tend to give our own group the benefit of the doubt, but we don't extend this courtesy to other groups.

And so it is that we trust our own group's intentions but doubt the intentions of other groups. For example, despite having a large arsenal of nuclear weapons, Americans "know" they would never use them except in self-defense. Indeed, to Americans, it is inherently obvious that their nuclear arsenal was built for deterrence rather than aggression. But Americans do not extend this trust to other countries. To Americans, it is also inherently obvious that Iran has no need for nuclear weapons other than to play a disruptive role in the Middle East. These "facts" are so obvious to Americans that when other countries deny their own aggressive intent, or question American claims, their denial and questioning are perceived as disingenuous and a strategic bargaining stance. Of course, from the perspective of America's rivals, the story is a mirror image, and such American claims are highly suspect.

* When we believe our own rationalizations, they are a form of self-deception.

In summary, humans evolved to be highly cooperative, but the underlying evolutionary pressure was a fight for survival. Our cooperative nature evolved to make us fiercer competitors. So, it should come as no surprise that our cooperative nature doesn't extend automatically to members of other groups. Indeed, other groups were often a serious threat to our ancestors' survival, so cooperation across group boundaries relies on a very tenuous form of trust. Due to the fact that fairness is relative, this trust between groups is threatened whenever we perceive another group as benefiting more from an agreement than we do, even if that agreement is clearly more beneficial than no agreement at all. Lastly, because we hypocritically perceive the motives of our own group positively but the same motives of other groups as suspect, we are unwilling to extend the benefit of the doubt to other groups but are dubious when other groups fail to extend the benefit of the doubt to us.

All these psychological factors pose important obstacles to our ability and willingness to achieve lasting peace and security with members of other ethnic groups, religions, and nations. But our evolved psychology is also highly sensitive to context, as it is the flexibility of human cognition and behavior that has made us such an evolutionary success story. Thus, the barriers to peace imposed by these deeply ingrained psychological tendencies can be overcome, not through reassurance or denial, but through structures, processes, and agreements that align the interests of previously hostile groups or through agreements and verification strategies that bypass these concerns.

When people see their interests aligned with other groups, or when they believe that it is impossible to cheat on an agreement, they are no longer hypervigilant for signs of cheating; nor are they tempted to cheat themselves. It can be difficult to align the interests of different groups, particularly when the groups have a long history of conflict, but numerous societal forces can achieve this goal over time. Increasing democracy, greater awareness and understanding of other cultures through increased contact, and numerous other societal changes drive groups toward cooperation and away from conflict.

The increasing integration of the international community through travel, trade, and tourism (as well as increasing integration over the internet) has the potential to create in many people an overarching group identity as fellow humans, rather than as members of certain tribes, ethnicities, countries, or religions. When such alignment of interests and redrawing of group boundaries are not possible, scientific advances in the capacity to detect cheaters, combined with agreements that are based on a realistic understanding of our evolved psychology, create the circumstances that make it possible to trust by verifying.

Part III

Using Knowledge of the Past to Build a Better Future

9

Why Evolution Gave Us Happiness

One Friday morning in 2007, commuters on the Washington, DC, Metro got a once-in-a-lifetime opportunity. As a study in human psychology, the *Washington Post* arranged for Joshua Bell, one of the world's great violinists, to busk in a downtown Metro station. For nearly forty-five minutes, Bell played classical music on his Stradivarius, during which time more than a thousand commuters entered the doorway directly in front of him. The *Post* had made elaborate plans in case the crowd got out of hand, but its preparations proved unnecessary. Only seven people stopped to listen to Bell for more than a minute.

What these commuters didn't know was that thousands of people pay a small fortune to hear Bell play in symphonies all over the world, at which point they have to dress up in their finest, struggle to find parking, and sit at a great distance from the master. Much has been made of this demonstration and its outcome. How could the same man be so sought after yet so readily ignored? As the *Washington Post* explained, Bell was "art without a frame"—in the context of a Metro station, people didn't know what to make of his music and were unable to appreciate it. No doubt there is some truth to this interpretation, as the price of a painting goes through the roof when we discover it's a Picasso and drops through the floor when we discover it's not, even though the painting itself hasn't changed an iota. But there is more to this story than context. Intentionally or not, when it set up this event, the *Post* tapped into a deeply ingrained feature of the human psyche.

By way of explanation, consider a classic psychology experiment from the early 1970s. John Darley and Dan Batson of Princeton University were pondering the story of the Good Samaritan, and wondering why so many people fail to help the unfortunate soul who has been beaten and robbed. Jesus tells this parable to highlight that everyone is our neighbor, and even the lowliest member of society can play an important role. (Samaritans were a disliked group at the time.) Darley and Batson took away a different lesson, wondering if perhaps the Samaritan was the only one to help because there was nowhere else he had to be. The Levite and the priest were more important than the Samaritan, and the fact that they walked by the man in need raised the possibility that they had other demands on their time. So, Darley and Batson set out to test whether being in a rush might predict who helps and who doesn't. The two researchers are both incredibly nice people, but to drive home their point, they designed a study with a particularly sadistic streak.

In their experiment, Darley and Batson asked seminary students to give a brief talk about what can be learned from the parable of the Good Samaritan. The students were told one of three things: (1) they had plenty of time to go to the office where their speech would be recorded, (2) they had just enough time to get there, or (3) they would need to hurry because they were behind schedule. The students then headed off to give their talk, and on their way, they encountered a person in need of help.

Darley and Batson had paid an actor to lie on the ground moaning, positioned so the seminary students almost needed to step over him on their way to explain how important it is to help people in need. The critical question was how many of them would heed the advice they were about to give. Consistent with Darley and Batson's predictions, the students were much less likely to help if they were behind schedule than if they had plenty of time. But perhaps the most remarkable finding of all is that across the three conditions, only 53 percent of the seminary students stopped even to ask the man if he was all right.

The extraordinary failure of these students to help the needy individual gives us a clue about why commuters on the Washington Metro failed to stop when they walked past Joshua Bell: in both cases, people were focused on the future. I suspect that many of the seminary students didn't really notice the person in need of help. They clearly saw him, as some of them stepped right over him, but they were so busy thinking about how they could persuade other people to be helpful that they probably paid him almost no attention. In the same manner, the commuters in the Washington Metro probably barely heard Joshua Bell through the buzz of their own thoughts, as they focused on how to deal with their difficult boss or their coworker who keeps stealing their lunch from the office fridge.

As I discuss in chapter 6, this capacity to travel in time mentally and make complex plans for the future has given us an enormous selective advantage. Unfortunately, that advantage comes at a cost, given that the time we spend living in the future distracts us from the present. As a consequence, people often fail to appreciate the pleasures (or demands) of the moment because they pay so little attention to the here and now. In my own case, I can't tell you how many times I've barely tasted a delicious snack because my mind was on an upcoming lecture, our next vacation, or how I was going to explain yet another speeding ticket to my wife.

Our proclivity to live in the future and ignore the present is not an easy problem to solve, although the varied approaches to mindfulness around the globe reflect the fact that many people try. Most meditation practices teach people to live in the moment. This is a laudable goal, but it's incredibly difficult to achieve because it's at odds with an evolved skill that has served us so well over the last million-plus years. We have a great deal of difficulty shutting down thoughts of the future unless the demands or pleasures of the moment are so substantial that they drag us back to the here and now.

My dogs, in contrast, show no signs of this inner struggle. They live in the moment because they are incapable of casting their minds forward. Every treat I give them is devoured with gusto, regardless of whether it means we just finished dinner or are off to the vet. Of course, planning for tomorrow isn't their strong suit, and thus their lives are under my control rather than the other way around. As in so many other ways, evolution gives with one hand but takes with the other. And this, in turn, brings us to what may be the most important question of all . . .

Why Aren't We Always Happy?

I've often wondered what it would be like to win the lottery and suddenly have more money than I could ever reasonably spend. I won't win because I don't play, but of course most people's lottery dreams never come true. This is not such a bad thing. As hard as it is to believe, lottery winners are usually no happier than they were before they won, and a fair few of them are a lot *less* happy. Not the day after they win—that's a pretty good day—but by a year or two later, most people have adapted to their new normal, and their happiness has returned to where it was before they drew the winning ticket. They may be driving a nicer car, but their mind is focused on the fact that they're still sitting in traffic.

Worse yet, some of them are focused on all the problems that their windfall has brought them, as friends and family come out of the woodwork expecting them to share their good fortune. As Sandra Hayes put it after winning the $224 million Missouri lottery in 2006, "I had to endure the greed and the need that people have. . . . These are people who you've loved, and they're turning into vampires trying to suck the life out of me."

The sad truth is that all of us have dreams, but even when our dreams come true, we rarely end up happier than we were before. New successes bring new challenges. The German folk saying *Vorfreude ist die schönste Freude* ("Anticipated joy is the greatest joy") is much more accurate than Disney's "happily ever after."

Why did evolution play this dirty trick on us, giving us dreams of achievements that will provide lifelong happiness but then failing to deliver the emotional goods when we achieve our goals? Some have blamed our modern world and the many discrepancies between our current lives and how we used to live (more on this later), but there is more to it than that. The advent of agriculture has led to major changes, many of which are disruptive for happiness, but our hunter-gatherer ancestors were also incapable of achieving lasting happiness.

The more important answer to this question lies in the fact that evolution doesn't care if we're happy, so long as we're reproductively successful. Happiness is a tool that evolution uses to incentivize us to do what is in our genes' best interest. If we were capable of experiencing lasting happiness, evolution would lose one of its best tools.

By way of example, consider two hypothetical ancestors, Thag and Crag. Both of them are sitting in a cave during the Pleistocene, eating lizard tails and dreaming of killing a mastodon. It's a pretty big ask to slog across the freezing glacier only to face such a dangerous animal, but in our hypothetical scenario, both of them live the dream and single-handedly kill the beast—actually, that's too unlikely; let's put them in charge of the hunt. As expected, both are incredibly happy and are the toast of their respective clans.

Imagine what would happen if Thag stayed incredibly happy forever while Crag dropped back to baseline within a week or so. Thag no longer feels the need to go out and kill anything, as he is content to relax in the cave and relive the exploits of his hunt. Crag, on the other hand, is motivationally hungry again, with a need to achieve. His continued ambition will get him off his duff and back out on the ice. This will result in further successes, which will attract a mate and the respect of his clan, and maybe his appreciative friends will see to it that he gets to sleep a little closer to the fire.

But our happy hippie Thag will soon be of little interest to the group by virtue of his lack of productivity. No one will want to hear his story about the mastodon anymore, and people will start asking the age-old question, "What have you done for me lately?" He won't particularly care—he is, after all, permanently happy—but he'll suffer the social and reproductive consequences regardless, and there will be fewer baby Thags in the next generation. As is evident in this epic tale of Thag and Crag, our inability to achieve lasting happiness pushed our ancestors to reach for further goals, which in turn meant that they left more offspring in the next generation.

We see a similar pattern today when we examine the motivational effects of happiness over time. Really happy people are rarely high achievers because they simply don't need to be. As Ted Turner put it, "You'll hardly ever find a super-achiever anywhere who isn't motivated at least partially by a sense of insecurity." The data agree with Ted. Consider the relationship between prior happiness and future earnings documented by Shigehiro Oishi of the University of Virginia and his colleagues in Figure 9.1.

On the far left of the graph we see that people who were unhappy in the mid-1980s (charted on the x-axis) go on to earn less money in the early 2000s (charted on the y-axis) than their happier compatriots. No surprise here: happy people are more energetic and compelling than sad people, and being energetic and compelling helps them earn more money.

More important for the current point, the moderately happy folks who sit in the middle right of this graph went on to earn the

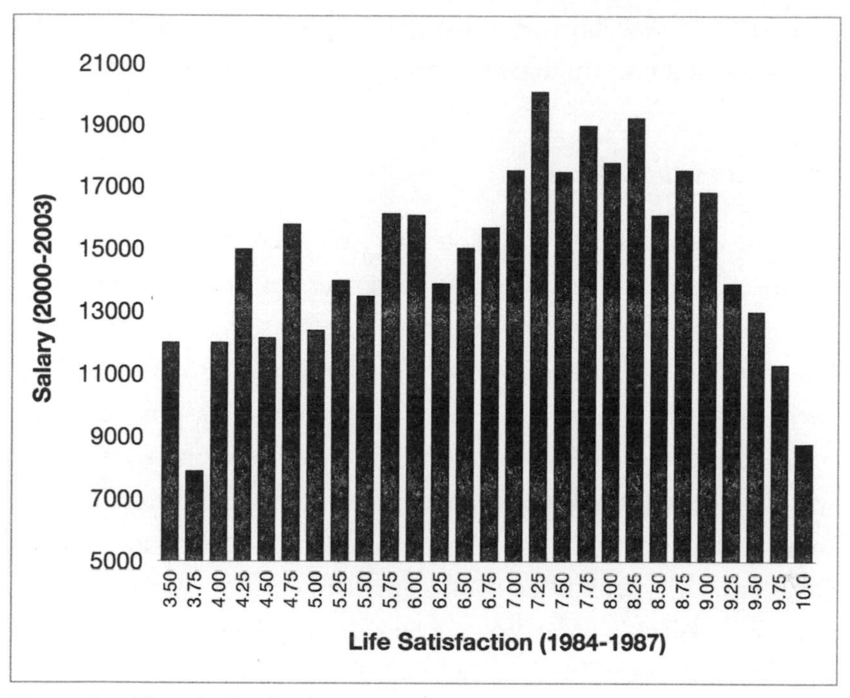

Figure 9.1. The relationship between earlier happiness and later income.
(Adapted from Oishi et al., 2007)

most money fifteen years later, while the earnings of the very happy folks on the far right look a lot like those of the unhappy ones. Some joy is clearly good for success in life, but too much happiness is a financial disaster. This is why evolution designed us to be reasonably happy, with occasional moments of giddiness that soon fade as we return to our individual baseline level of happiness. Numerous self-help professionals would have us believe that attaining maximal or permanent happiness should be our goal, but an evolutionary perspective clarifies that such a goal is neither achievable nor desirable.

Happiness evolved for a reason—it gets us out there killing mastodons. But happiness is more than just a motivator; it also plays a critical role in the connection between mind and body. So, before turning to the question of exactly what makes us happy, let's spend a little time figuring out why happiness is important even for the unrepentant curmudgeons among us.

Happiness and Health

Late one evening about a decade ago, I received an unexpected call on my cell phone. The caller was from overseas, and the connection was poor, so I missed his name but caught something about his being in an anthropology department at what sounded like Rutgers University. I didn't ask him to repeat himself, as I figured that most people don't call at that hour and it would soon become clear who he was and what he wanted. After a bit of idle chitchat, the caller told me that he had been invited to take a sabbatical in Berlin the following year, and he wanted me to join him. I have nothing against sabbaticals or Berlin—in fact, I'm a big fan of both—but I have my own commitments, so I was politely excusing myself from this unusual request when it dawned on me that I was talking to Robert Trivers.

I'd had the good fortune of meeting him at a small conference about five years prior to that phone call, and although I'd spoken with him only a few times in the intervening years, he has a distinct, gravelly voice that I eventually recognized over the poor telephone line. If it was Trivers who was inviting me on sabbatical, then that changed everything, so I told him to hold on for a moment while I covered the mouthpiece and asked my wife how she felt about a sabbatical in Berlin. She thought that sounded like fun, so I told him to count me in.

We arrived at the wonderful Wissenschaftskolleg in Berlin in October 2008. The plan was for me to spend the next six months working with Trivers to help develop his theory of self-deception (discussed in chapter 5). Trivers is legendary not just for his brilliance, but also for his mercurial temperament—you don't need to take my word for it; just get a copy of his autobiography, *Wild*

Life—and at that point I didn't know him well. So, I was a little concerned about how our collaboration would proceed. It turned out that we got along famously, but that first week, we definitely got off to a rocky start.

In one of our first meetings, Trivers suggested that our immune system can serve as a bank—I had no idea what he meant by that, but I nodded thoughtfully—and that we evolved to orient ourselves toward positive thoughts as we aged in an effort to enhance our immune functioning. (I wasn't sure what he meant by that, either.) I was aware that older adults tend to remember positive things in life better than negative things, whereas younger adults remember the positive and negative equally well. The predominant theory in psychology to explain this effect is that older adults are aware that they have a limited time left on this planet, so they prioritize positive emotional experiences (much as students do when they are about to graduate and leave their friends behind). I had always found this theory convincing, and felt that with his alternative theory, Trivers was barking up the wrong tree. I was also aware that evolution has a much stronger impact on people before they reproduce than after, so an evolutionary basis for the increase in positivity with age seemed unlikely to me.

After a fair bit of arguing (and a fair bit of explaining), Bob eventually convinced me that this was an idea worth testing. As I discuss in the prologue, in our premedical past, grandparents played an important role in the survival of their grandchildren (more on this later), which suggests that their longevity is important to their grandchildren. Indeed, humans show several adaptations that enhance functioning late in life, such as changes in our genetic makeup that protect us against Alzheimer's and other forms of neurological decline that impact only older adults. These findings raise the possibility that evolution might also shape older adults' motivational system to keep them alive as long as possible. But what did Trivers mean with his talk of the immune system as a bank, why would older adults capitalize on it, and what does any of this have to do with the importance of happiness and its relationship to health? Let me answer these questions one at a time.

When we consider the costs of development and maintenance, our brain is our most expensive organ. It demands 20 percent of our metabolic energy at all times, whether we are solving math problems or watching reruns. Due to its constant energy requirement, there is no way to borrow energy from the brain when our need for energy exceeds the available supply. In contrast, our muscles use way more energy when they are active than when they are at rest, so in principle, we could borrow energy from our muscles by sitting down and relaxing. The problem with this strategy is that most of the energy-demanding emergencies our ancestors faced required a muscular response. There was no way to borrow energy from our muscles during an emergency because sitting down when a mastodon showed up was not an effective option.

This brings us to our immune system, which also hums along at great metabolic cost, but largely in the service of future needs rather than current ones. At any one moment, we have an enormous number of immune cells coursing through our body, so we can afford a momentary break from production if the demands on our metabolic energy supply become pressing. When our body is in need of more energy than it has, one of the first things it sacrifices is immune function.

When did our ancestors face pressing energy demands? Running from hungry saber-toothed tigers and thwacking enemies with a club would be prime examples. When you can feel a tiger's hot breath on your neck as you cover the last few yards to the nearest tree, you don't really need to waste energy making more immune cells to fight off tomorrow's cold. What you do need is to shift all available resources to your legs, with the hope that you'll live to see another cough or sneeze. Because fighting and running for our lives are not happy times, evolution took advantage of this fact to link our bodily systems to our psychological states.

As a result, our immune system evolved to hum along at peak capacity when we're happy but to slow down dramatically when we're not. This is why long-term unhappiness can literally kill you through its immune-suppressing effects, and why loneliness in late adulthood is deadlier than smoking. Indeed, once you're over sixty-five, you're better off smoking, drinking, or overeating with your friends than you are sitting at home alone.

With this background in mind, Trivers hypothesized that older adults evolved a strategy of turning this relationship on its head, becoming more focused on the positive things in life in an effort to enhance their immune functioning. Such a strategy would be more sensible for older than younger adults for two reasons. First, older adults have a weaker immune system than younger adults, and face greater threats from tumors and pathogens. Second, older adults know much more about the world than younger adults do, so they don't need to pay as much attention to what's going on around them. For example, when older adults interact with a surly bank employee or a harried flight attendant, they have a library of related experiences to draw upon and can respond to the situation effectively without giving it much thought. As a result, they can afford to gloss over some of the unpleasant things in life.

When I returned from Berlin to my lab at the University of Queensland, I raised this hypothesis with my collaborator Julie Henry, and with Trivers on board we asked my student Elise Kalokerinos if she would like to take the lead in testing the idea for her PhD. Elise thought it sounded like fun, so she set about finding a way to test it. Over the course of the next year, she brought young and old adults into the laboratory and showed them photographs of nice things, such as baskets of puppies, and photographs of nasty things, such as plane crashes, and then tested their memory for the pictures. Sure enough, our participants who were over sixty-five tended to remember the puppies better than the plane crashes (which suggests that they were paying more attention to the positive), while our younger participants remembered both equally well.

Elise then asked our older participants to return to the lab one and two years later so we could draw blood to assess their immune functioning. The immune system is vast, but in this initial study, we decided to focus on a class of white blood cells known as $CD4^+$s. These cells facilitate immune functioning by triggering other white blood cells (known as B cells) to produce antibodies. Elise found that better memory for positive but not negative pictures was associated with higher $CD4^+$ counts and lower $CD4^+$ activation* one and two years later.

Higher $CD4^+$ counts usually indicate a greater preparedness to fight off illness. In contrast, higher activation of $CD4^+$s indicates that the person is busy fighting infection, and thus is in poor health. In other words, the more positive their memories were today, the healthier they were next year and the year after. This relationship between positivity in memory and $CD4^+$s raises the possibility that by focusing on the positive aspects of life, we enhance our own immune functioning.

These findings are not a good fit with the theory that positivity with age is caused by older adults' awareness of their limited time on this planet; but they are consistent with other research showing that happiness plays an important role in health and longevity. For example, when researchers intentionally expose people to cold viruses, they find that people who are happy and have good social support are less likely to catch a cold than unhappy people and those with poor social support. Happy and well-supported people also heal more quickly when intentionally wounded in the name of science.

* Activated $CD4^+$s are busy fighting infection; prior to activation, they are simply available if needed.

This effect holds for our primate cousins as well. Wild monkeys in the mountains of Morocco who have stronger friendship ties show a decreased physiological stress response (i.e., reduced steroid hormones in their feces) to cold weather and aggression from other monkeys. Notice that the key issues for monkeys and for us are friendship and social support. Satisfying relationships play an important role in proper immune functioning.

My favorite experiment on this effect is by Jan Kiecolt-Glaser at Ohio State University and her colleagues. In their study, they brought couples into the lab on two different occasions to create blisters on the inside of their forearms by attaching small vacuum suction devices. After they had suctioned up eight blisters on the participants' arms, they then snipped the skin off the top of the blisters and put little plastic tubs over them. (I know this sounds horrific, but apparently, it's not as bad as it seems.) These tubs were now artificial blister chambers that the researchers could use to collect blister goo to examine cellular immune responses throughout the experiment.

On the first visit to the lab the researchers asked the couples to discuss their relationship history, and on the second visit they asked them to discuss areas of ongoing conflict (such as money or in-laws). The researchers made themselves scarce while these conversations took place, but recorded them for later analysis. Despite the semi-public nature of the setting, couples were quite willing to speak their minds (e.g., "You're only being nice so I'll have sex with you tonight," "You were being mean on purpose").

This experiment yielded several interesting findings. First, the blisters took one more day to heal after the conflict discussions than after the more positive initial conversations. Second, the blisters took an extra *two* days to heal among couples who were hostile to each other during their conflict discussions compared to the couples who were not. Third, the cellular immune activity inside the blisters mirrored what was going on with the healing. There were large increases in inflammation following hostile conflict discussions but almost no increase in inflammation following non-hostile conflict discussions.

These inflammatory responses are implicated in cardiovascular disease, diabetes, and other illnesses, which is precisely why happy marriages help us live longer and why unhappy marriages (and loneliness) shorten our lives. These findings also clarify why even people who enjoy living alone need regular human contact, a group that matters to them, and meaningful friendships. People differ in how many friends they need and how often they need to see them, but everyone requires social connections to maintain health and happiness.

So, what is the purpose of happiness? As you can see, there is no single answer to this question. Happiness motivates us to do things that help us survive and reproduce, but happiness is not an end in and of itself. Evolution often sacrifices our happiness in the service of other goals; those who cannot feel pain and despair are severely constrained in their ability to learn to avoid bad people, situations, and ideas. Indeed, negative emotions are just as important as positive ones (perhaps even more so), as the costs of plans gone awry can far outweigh the benefits of success.

We saw in chapter 5 that happiness is readily apparent to others and its signaling value is not lost on people who are trying to size you up as a potential partner, ally, or enemy. In this sense, happiness is also a critically important social emotion. Happiness serves as a signal to our own bodies as well, by communicating that now would be a good time to expend energy on repair and illness prevention.

10

Finding Happiness in Evolutionary Imperatives

Now that we've discussed the purpose of happiness, we can turn to the question of exactly *what* makes us happy. Because evolution shaped our motivational system, consideration of our evolutionary imperatives should give us some traction on how to lead a good life.

Given our particular imperatives, you may think an evolutionary guide to happiness would be a pretty short pamphlet, or perhaps even just the simple equation:

$$F^{ood} + S^{ex} = H^{appiness}$$

No doubt there is some truth to this equation, but the story is also much more complex. Happiness may be a matter of meeting our evolutionary imperatives, but these imperatives are often at cross-purposes with one another, and a good deal of wisdom and self-knowledge is required to navigate successfully among them. This challenge can be seen most readily in our two paramount goals of reproduction and survival. Despite the fact that we must achieve both to pass our genes on to the next generation, reproduction is the currency of evolution, and survival is important only to the degree that it serves that goal.

The challenges inherent in managing our evolutionary urges are magnified by numerous other complexities, some ancient and some modern. On the ancient side, although our imperatives are universal, our strategies for achieving them are not. For many species there is only one way to get ahead in life. If you are a male dung beetle, you must be able to roll a ball of feces that is many times your body weight. If you are a male moose, you must be able to knock other moose on their bums when you butt heads. But for humans, the number of strategies available to us is limited only by our imagination. Do I curl, straighten, or cut my hair? Do I whiten my teeth or spend my money on braces? Do I pose for my Tinder photo next to my stamp collection or my boxing trophy? To understand how to achieve happiness, we need to understand how our personalities, proclivities, and abilities *each* can make us a success. Not every road leads to happiness, but different roads suit different people.

On the modern side, our highly technological world continues to develop new tricks to short-circuit our pursuit of happiness via what Robert Trivers calls phenotypic indulgences. Phenotypic indulgences, while pleasing, are really just surrogates for our evolved preferences. Alcohol and drugs, television, and even potato chips are all phenotypic indulgences. They mimic ancient pleasures without delivering the outcomes that made those ancient activities adaptive and hence pleasurable. (Think of the difference between watching *Friends* and having friends.)

If the notion that humans have evolutionary imperatives strikes you as overly deterministic, it is important to remember that we also evolved to be the most cognitively flexible species on the planet. As I discuss in chapter 2, humans need to learn an extraordinary amount of information to survive and thrive, and more than any other animal we chart our own life course. That doesn't mean we can find happiness in everything we do—most of us cannot—but it *does* mean we get to decide the importance of happiness in our life as well as the most fruitful way to pursue it. Understanding human nature through the pressures exerted by our evolutionary past can guide us in this pursuit and also help us understand happiness itself.

An Evolutionary Guide to Happiness

Evolution depends on reproduction, first and foremost. But that fact has led to two common misconceptions. The first misconception is the belief that we must have children of our own if we are to pass on our genes. In fact, we can also be an evolutionary success if we enhance the reproductive success of our relatives. Evolutionary pressures can lead people to be a good uncle or auntie in the same manner as a good parent, and the outcome is much the same. Helping nieces and nephews ensures that copies of the good uncle's or auntie's genes will have a better chance of making it into the next generation.

The second misconception is the widespread belief that evolution gave us a desire to reproduce. Just because evolution relies on reproduction doesn't mean that humans evolved a desire to have children. Until very recently in our evolutionary history, we had no idea that sexual relations created children, so evolution would have achieved nothing by giving us a desire for children. Rather, evolution gave us a strong desire to have sex, and then (because we are a species that requires parental care) tossed in a tendency to feel fond of any children we produced. By evolving sexual desire plus nurturance, we arrived at the same outcome as we would have if we'd evolved to want children and knew how to produce them. This combination of wanting sex and feeling nurturant toward resulting offspring works for us and every other mammal (or at least every other female mammal; biparental care is pretty rare among our furry cousins).

You might argue that a desire for sex in the absence of a desire for children is inefficient, and indeed it is. Humans (and other primates) have all sorts of "wasteful" sex that cannot possibly lead to reproduction. I suspect that the evolution of hands was followed forty-five minutes later by the invention of masturbation, which would have been impossible with hooves and too risky with claws. Nevertheless, so long as we engage in enough of the reproductive sort of sexual activity, the wasted energy cost of nonreproductive sex is likely to be low.

Due to the central role of sex in reproduction, frequent sexual activity (particularly if you can arrange it with someone you like) is a key to the good life. But frequent sexual activity alone is insufficient to successfully reproduce, and hence insufficient for a happy life. The long period of dependency in human children dictates that parenthood is also critically important for reproduction. In fact, children are so difficult to raise that parents are barely sufficient, and thus evolution invented the grandparent.

As I discuss in the prologue, Lahdenperä and her colleagues found that our ancestors were more likely to survive childhood and mothers were more likely to have children in rapid succession if they had the help of a grandmother. How did evolution create grandmothers? By preventing women from producing more children of their own while they still had plenty of life in them, evolution gave them the opportunity to focus on their grandchildren rather than their children.* This is why human females evolved menopause.

As can be seen in Figure 10.1, when Susan Alberts of Duke University and her colleagues compared humans with other primates, they found that we are unique among our primate cousins in the female tendency to outlive fertility. The other apes and monkeys on this graph all fall along the diagonal line, indicating that they remain fertile until they die. For example, the latest that chimp

* Remember, women had no control over their fertility at that time.

females tend to give birth is around age forty, and the longest they tend to live is also around age forty. The human example is from !Kung hunter-gatherers, who have little or no access to modern medical care and thus give us a better sense of how our ancestors lived than we could get from humans living in industrialized countries. Here we see that the latest age at which !Kung women tend to give birth is in their early forties, but the longest they tend to live is into their mid-seventies. If there were no need for grandparenting, it would be an unusual evolutionary arrangement for women to outlive their fertility by such a wide margin. (Remember, evolution doesn't care about survival for its own sake.)

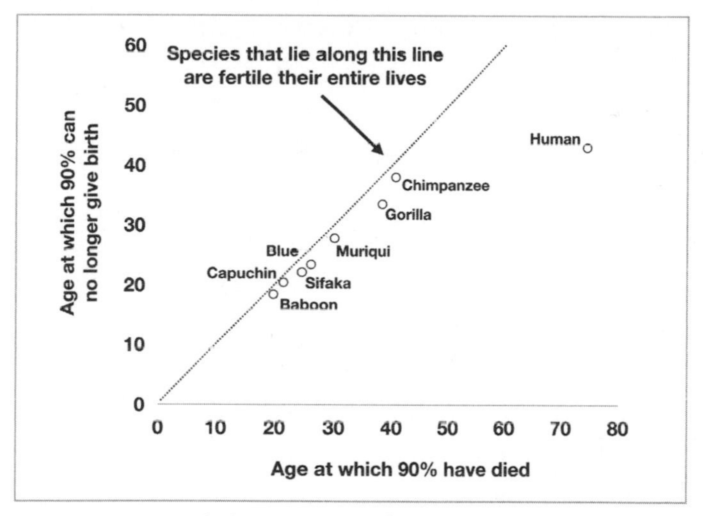

Figure 10.1. Female longevity and fertility in various primates. (Alberts et al., 2013)

Raising and teaching our children and grandchildren are thus an important source of life satisfaction. That doesn't mean that we evolved to enjoy every moment we spend with them—when my kids were little, Monday morning and off-to-school-you-go couldn't come soon enough—but it does mean that we get enormous satisfaction from seeing our children succeed in life. Just a few minutes at your child's high school graduation or wedding provide ample demonstration of that fact.

This next point might sound sexist, but as I discuss in chapter 4, women have a much greater obligatory biological investment than men in creating and raising children. As a consequence, it is likely that raising children (and grandchildren) plays a larger role in female than in male life satisfaction. Be that as it may, it is in both men's and women's interest to facilitate the survival of their offspring and those of their close family, so providing for the next generation of kin is likely to be an important source of happiness for everyone. This is not to say that the day-to-day tasks of raising children are fun, as they often aren't, but the knowledge that you've done the right thing for your children is a great source of life satisfaction.

So far this recipe for happiness, have sex and do a good job raising your children, probably seems inherently obvious to the most casual observer. But reproduction is more complicated than that, and so are its implications for life satisfaction. One of the most complicated aspects of human reproduction is finding the right partner in the first place. Choosing a long-term partner requires an ability to predict your preferences far into the future, and you need only consider your prior clothes or hairstyle to realize how difficult that task is. Indeed, I struggle at the fruit market each week trying to predict which bunch of bananas will ripen before they turn brown, a decidedly easier task than predicting if Courtney or Kim will be more likely to interest me for the next forty to fifty years.

The difficulty of this prediction problem is exacerbated in a species like ours, which forms enduring pair bonds, as partnership is a *mutual* decision. If we were laughing tree frogs (a real Australian frog, by the way), then every female could mate with the most desirable male, as his role in the process involves nothing more than fertilizing the eggs. But because we work together to raise our young, it is not possible for every woman to mate with the most desirable man, or vice versa. Rather, the resultant compromises necessitated by limited availability and mutual choice demand that we trade some preferred aspects for others, and of course different people are likely to make different trade-offs when choosing a partner. You might care more about kindness, I might care more about shared values, and someone else might be particularly concerned with brains, beauty, or finances.

The best way to solve the problem of mutuality is to be as desirable as possible, as that increases the chances that the one you love will love you back. For this reason, evolution motivates us to do things that improve our chances of attracting and keeping the person with whom we most want to have sex and children. In other words, we try to be what the other sex is looking for.

What the other sex is looking for often feels like one of Life's Great Mysteries, but it's really not so mysterious at all. Men and women want many of the same things in a partner; kindness and generosity are near the top of everyone's list, and it doesn't hurt to be sexy, fun, and smart. But as I discuss in chapter 4, it's not enough to be smart or sexy. It's critical to be smart*er* or sexi*er* than the people around you, or you're still going to be the last one chosen. All our attributes are relative. That doesn't mean that you need to be tops in all domains, but it does mean that you need to stand out in those areas where you have the best prospects. In my own case, my towering five-foot-six stature combined with a nine-inch vertical leap meant that my prospects were poor in basketball, so I never put much effort into the game. But my prospects were better in tennis, and I put in a fair bit of time on the court in a (failed) effort at self-improvement.

It is important to keep in mind that I was never pursuing excellence on the court to get the girl (although the high school letter jacket and all that it entailed might have crossed my mind). I believed that I played tennis because I loved the game. Ultimately, though, it doesn't really matter what we think our motivation is.

What matters is the consequences of our actions. If being a great tennis player is attractive to others, then my "love of the game" would have evolved because it enhanced my reproductive success.

Similarly, the drive to be better than those around us often emerges in a desire for mastery. Mastery is important because our unique skill set differentiates us from others and makes us desirable as a mate. But the pursuit of mastery can be costly, as the underlying concern with relative status can easily put us on a hedonic treadmill, with each accomplishment quickly fading as we strive to keep up with (by which I mean surpass) the Joneses. Like our ancestor Crag, once we beat the Joneses, we set our sights on the Smiths. As a consequence, markers of success such as wealth have only a trivial effect on happiness unless we have more than those around us, suggesting that it is status and not money that we are really after. Two sets of findings illustrate this point nicely.

With regard to status, research on monkeys demonstrates that when they rise to the top of the status hierarchy, there is an increase in the dopamine (evolution's pleasure drug) sensitivity in their brains. As a result of this increased dopamine sensitivity, monkeys at the top of the heap no longer enjoy cocaine (a drug that hijacks the dopamine system). When offered cocaine versus salt water, these top monkeys show no preference between them. In contrast, monkeys at the bottom of the status hierarchy have low dopamine sensitivity and become avid coke users. Data such as these confirm the common wisdom that high status makes us happy and low status makes us sad.

With regard to money, once people get out of poverty, the relationship between wealth and happiness is not as strong as you might think. Much more important, if all of society rises in wealth at the same time, increases in wealth beyond poverty provide *no* increase in happiness. Zippo. This effect can be seen by plotting life satisfaction against purchasing power over the last fifty-five years in the United States (Figure 10.2). As is evident in this graph, dramatic society-wide increases in real wealth (i.e., controlling for inflation) led to no associated increases in happiness.

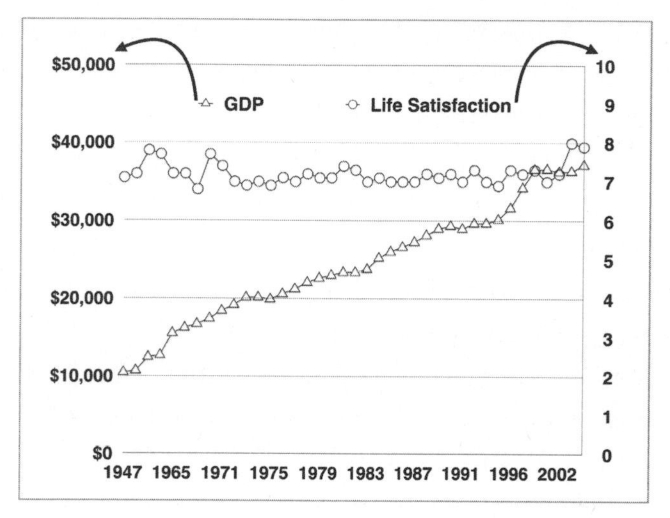

Figure 10.2. Real income (controlling for inflation) and life satisfaction in the United States.

These data suggest that my home cinema, granite countertops, and convertible don't actually make me any happier unless I have them and you don't. In other words, I want these things only to put myself above others. Moreover, *whether I know it or not*, the reason I want to rise to the top of the heap is because that gives me a better chance of getting the partner I really want. The TV, countertops, and car are just trivialities, but because I don't know this, I spend my time coveting them, working to acquire them, and eventually becoming the disinterested owner of them.*

Unfortunately, getting off this hedonic treadmill is no easy matter. Millions of years of sexual selection have etched status concerns into the deepest levels of our psyche, so turning them off or even

* I'm not saying we have a strong desire to stand out, because we don't. We very much want to fit in, but we want to fit in at the top of our group, not the bottom.

ignoring them is impossible for most of us. But awareness of the problem probably helps, particularly since it can allow us to focus our attention on other aspects of our lives that have the potential to provide more lasting happiness. Awareness can also help us be better parents or friends. We may be unaware of our own underlying status strivings when we lunch with the boss or try to improve our tennis game, but we are gifted with the ability to see through others. When we see our friends and family worrying about how they stack up against everyone else, we can be a bit more sympathetic to their insecurities rather than asking why they care so much about what other people are getting and doing.

Ignoring our ingrained status concerns may be impossible, but one solution is to spend money on activities rather than material goods—buy things to do rather than to have. If you're like me, this possibility will strike you as counterintuitive, mostly because expensive experiences seem so self-indulgent. I remember feeling pangs of guilt in graduate school when I decided to spend my savings on a ski vacation despite the fact that my only couch had been rescued from the side of the road. Even as I left for the airport, I asked myself if furniture might have been a wiser and longer-lasting purchase than a trip to Aspen. But it turns out that I had it exactly backward: the ski trip was a longer-lasting purchase than the couch. My ski trip from 1987 still makes me happy when I think about it today, my friends and I still talk about what a great time we had, and my wife would have long since tossed out the Naugahyde sectional I was considering in place of that trip.

Research by Leaf Van Boven and Thomas Gilovich of Cornell shows that I'm not alone in this regard. As you can see in Figure 10.3, particularly when people move from buying necessities to buying luxuries, their experiential purchases make them a lot happier than their material ones. This relationship holds even when the same object is bought for material reasons (I want to own a fancy

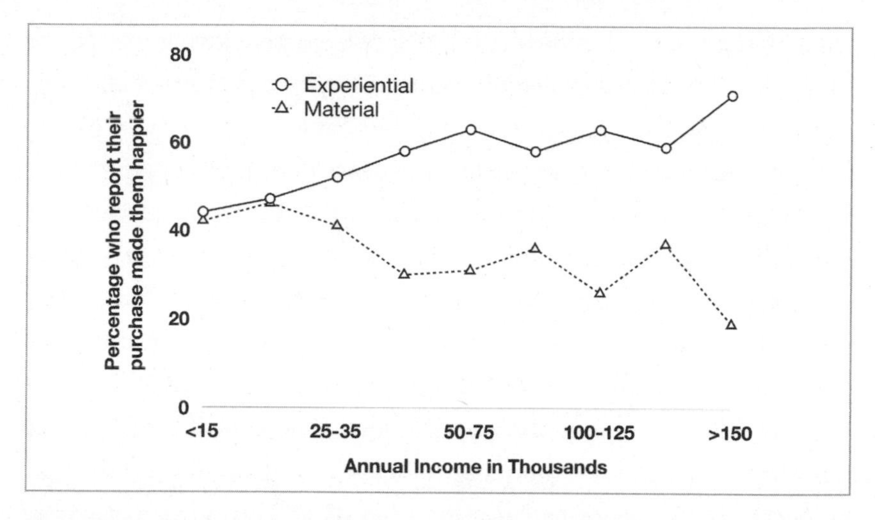

Figure 10.3. The happiness effects of material versus experiential purchases. (Adapted from Van Boven and Gilovich, 2003)

car) versus experiential reasons (I love to drive my new Jag down winding country roads).

The next time you have money burning a hole in your pocket, keep in mind that experiences are a better hedonic investment than material purchases. The things we own lose their allure as soon as we reset our status goals, but the things we do become a part of us. Positive experiences give us the stories we tell friends and family—our most important memories—and continue to provide satisfaction long after the experience has ended.

Happiness and Survival

Survival goals are basic to all living beings, and many of our emotional responses evolved due to their survival value. We enjoy eating fat, sugar, and salt because they were rare in our ancestral environment, but critical for survival. We feel anxious and afraid when we walk through the woods at night because our reliance on our eyes over our ears and nose means that we are far more likely to be prey than predator once darkness falls. We are hypersensitive to the possibility that our friends or neighbors might reject us because expulsion from our group was an existential threat for our ancestors. We gain comfort and security from home and hearth because they provided our ancestors protection from the elements and predators. All these preferences have been around far longer than we've been *Homo sapiens*.

However, despite their importance, survival goals are often trumped by reproduction goals. The most fundamental example of this trade-off can be seen in the aging process itself. We grow old and die largely because we spend precious biological resources on our efforts to attract, retain, and reproduce with mates rather than on tissue maintenance and repair. If we passed on our genetic material through longevity rather than reproduction, evolution would have ensured that we spent enough resources on tissue maintenance to enable us to live for centuries rather than years. In principle, such an outcome is possible, as longevity enables longer periods of reproduction and hence more offspring, but the prevalence of predators and parasites made any strategy based on longevity an unlikely prospect. Because our ancestors rarely had the opportunity to die of old age (recall Figure 8.1), efforts spent on living longer were largely wasted, and thus biological resources were better spent on more immediate mating goals. For this reason, a trait that helps us reproduce when we're young will typically have a selective advantage, even if it kills us when we're old.

Such an effect can be seen in the ε4 allele of the ApoE gene, which is associated with an increased likelihood of developing Alzheimer's late in life. Ironically, this allele is also associated with better cognitive functioning early in life. As a consequence of the benefits it brings when we're young, this killer gene is widespread in the population. Evolution's tendency to sacrifice survival in service of reproduction means that we have numerous self-destructive tendencies that might be considered the psychological equivalent of the ε4 allele of the ApoE gene.

Perhaps the most famous example of a psychological ε4 allele can be found in male risk taking and conflict. Often referred to as testosterone poisoning or just male stupidity, the biggest demographic risk factor in most industrialized societies is the combination of being young and male. Figure 10.4, based on research by biologist Ian Owens at Imperial College London, uses mortality data from the United States in the late 1990s to show that once men hit puberty, they are much more likely than women to die in conflicts with one another, car crashes, and pretty much every other type of accident.

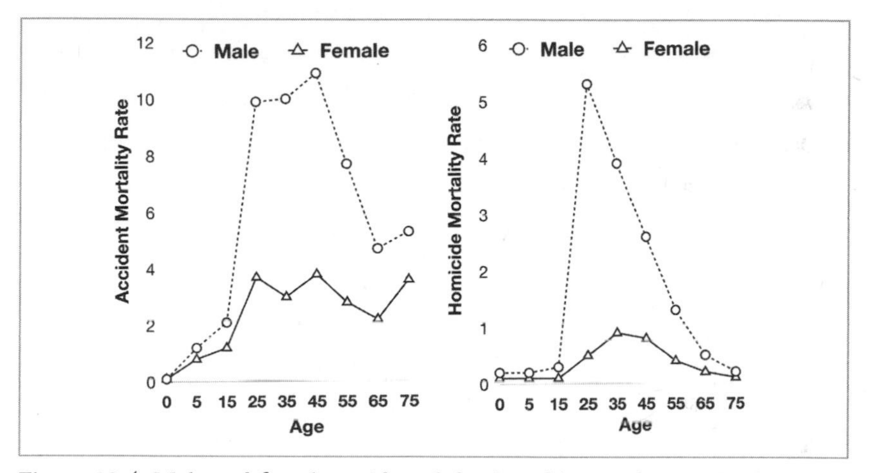

Figure 10.4. Male and female accidental death and homicide rates in the United States. The y-axis shows annual deaths per 100,000 people per year. (Adapted from Owens, 2002)

What many people don't realize is that male stupidity is actually an adaptation to female choosiness, as male risk taking and conflict seeking are products of sexual selection. This may seem counterintuitive, as women are often the first to point out that they're not attracted by male stupidity, so let me explain what I mean.

You can trace your male ancestry through your Y chromosome (or your father's Y chromosome if you're female); you can also trace your female ancestry via your mitochondrial DNA, which is inherited only from your mother and traces your female line. If you were to conduct these analyses in a large enough sample, you'd find that there were many more females than males in your ancestry. At first glance this imbalance seems impossible, as it takes two to tango. But, of course, some men have lots of partners and father many children; other men not so much. Men are far more likely than women to have a very large number of children, and also far more likely to get left out of the mating game altogether.

Risk taking provides an opportunity to shift from being a monk to a Casanova, so men have evolved a tendency to take risks if those risks might pay off in reproductive opportunities. Because men have greater variability in their reproductive success (i.e., many men bear no children, and some men bear a large number), they are more inclined to take risks than women. In contrast, risk taking is imprudent for women, as they are highly likely to have a similar number of offspring whether they play it safe or take outlandish risks.

Male stupidity is further enhanced by the fact that risk taking itself is an honest signal of male quality. Thinking back to chapter 4, you'll recall that this means that risk taking communicates reliable information about how robust and skilled a person is. If you succeed at the risk, then you're skilled. If you fail but survive, then you're robust. If you fail and don't survive, well, that's indicative, too. For this reason, women can use male risk taking as a sign of quality, suggesting that men are likely to evolve a tendency to take risks when they have the opportunity to attract a woman. This idea had been widely demonstrated in the animal kingdom, but my PhD student Richard Ronay and I were keen to see if we could find any evidence for it in humans.

When testing a new idea, we typically begin as inexpensively as possible, so in our initial lab studies, we asked men to pump up cyber balloons on a computer. Every pump earned them money, but every pump also increased the risk that the balloon would pop and they would lose everything. Consistent with predictions, we found that men pumped the e-balloons larger after seeing pictures of attractive women. Chances are you're thinking the same thing we were: that's a perfectly good start, but it's not very impressive evidence of risk taking.

For our next experiment we bought ourselves some "shock balls" on eBay. This delightful game involves an electrified ball that randomly lights up and simultaneously shocks whoever is holding it. We decided to use the shock balls to create a game similar to the one we used with the cyber balloons. Men would get paid for every second they held the ball, but if they held it for too long they would lose all their money and get shocked as well. We thought that men might hold the shock ball longer if they were exposed to attractive women. Sure enough, that effect emerged as well, but we remained underwhelmed with the magnitude of the risk taking involved in this task. Shock balls don't really put a premium on skill or robustness.

The problem we faced is that it's not easy to study serious risk taking in the lab, as it's unethical to put people in a situation in which they might genuinely get hurt. We struggled with this problem for a while before Richard came up with a great idea: why not study skateboarders? They are taking some pretty serious risks already, and all we have to do is show up with an attractive woman and see if that makes a difference.

So, we hired a beautiful research assistant and headed off to skateboard parks. In the first stage of the experiment, a male researcher approached a skateboarder and asked if he could film him making ten attempts at a trick that he was working on but hadn't yet mastered. In the second stage of the experiment, the same skateboarder was either approached by the male experimenter or by the attractive female we had hired, who asked to film the same ten tricks. After the skateboarders completed their second round of tricks, we took a saliva sample to measure their testosterone.

Just as we expected, testosterone went up in the presence of the female experimenter, and the higher the testosterone levels, the more risks the skateboarders took. As a consequence of their greater risk taking, they crashed more often but they successfully landed more tricks as well. Both outcomes serve a purpose as far as the skateboarders are concerned. Successful landings display skill, and crash landings show robustness. Sure enough, our participants shrugged off bleeding elbows and knees, and also declined our ethically mandated offer of helmets and kneepads at the onset of the experiment.

What can we infer about happiness from this conflict between survival and reproduction? The first lesson is that risk taking and other foolish things that young men do are not "pathologies," signs of their disconnection from the modern world, or other labels often provided by social commentators. Rather, they are evolved strategies that made perfect sense for our ancestors and probably continue to make reproductive sense today.

The second lesson is that trying to prevent our sons, brothers, or friends from taking unnecessary risks is a bit like pissing into the wind. Removing the opportunity for young men to engage in competition and risk taking is a bad idea, and likely to lead to unpleasant blowback. Young men feel millions of years of evolutionary pressure, emanating from their testicles, pushing them toward risk and competition. For this reason, the best bet is not to eliminate risk entirely, but to replace truly dangerous risk and conflict with more benign opportunities for thrill seeking and competition. Sports in which you can't get hurt at all are unlikely to fulfill such goals, but sports in which you won't get hurt too badly are a great substitute.

Cooperation and Competition

As I discuss in chapter 1, our enhanced capacity for cooperation was the key adaptation that allowed our ancestors to survive the move from the trees. Because we evolved to cooperate with one another, we also evolved a cheater-detection system and a strong emotional response to free riders. We all know the feelings of anger and righteous indignation when others take advantage of us. This evolved response explains why one of the continuing arguments surrounding our welfare system is whether recipients are lazy people making suckers of the rest of us or disadvantaged people who deserve our sympathy and help.

Our outrage and anger when we are cheated ensure that others in our group cooperate with us, but our emotions have also been shaped by the underlying cooperation goal itself. We don't really like people who cooperate with us only to reciprocate our earlier help, or to gain our later cooperation in return. Rather, we like people who are friendly, kind, and generous, who enjoy cooperating for cooperation's sake. That, of course, means that other people like (or dislike) *us* for the same reasons, which gave a substantial evolutionary advantage to our ancestors who genuinely enjoyed cooperating. This is why we often share resources with strangers who will never be able to help us in return.

Economists are sometimes surprised when people share with strangers, but this surprise emerges from a misunderstanding of our evolutionary history. It may seem that we're setting ourselves up to be suckers, but even though generous people do get exploited, in the long run, they win far more than they lose. Generous people are more popular than stingy or calculating people all over the world. When Hadza hunter-gatherers in Tanzania break camp and go off in different directions, the generous ones have lots of people who want to be with them, while the stingy ones are at constant risk of being left alone. When the Martu people of western Australia head off on morning hunts, the generous people are always chosen as partners even if they're not the best hunters, while the stingy ones are often forgotten and left behind. The same effects hold among the Quechua pastoralists in the Peruvian Andes, and every other people on earth for whom we have the necessary data.*

As a consequence of these evolutionary pressures, we evolved to be mindlessly helpful, automatic cooperators: we cooperate without thinking because cooperation is our default reaction when people need our help. When people participate in experiments in which they must choose whether to cooperate with others or defect, cooperation is chosen more quickly than defection, even if the sensible choice is to defect. The same effect emerges when people are forced to make snap decisions; they are far more likely to choose cooperation.

I remember being delighted when this research on mindless helpfulness was first published, as it explained my completely idiotic behavior when I risked a child's life for a forty-cent ice-cream cone. About a dozen years ago I was riding down an escalator with my wife and two small children while carrying a McDonald's ice-cream cone that I had just purchased for them. As we started the ride down, I heard a woman screaming at the bottom. When I looked toward her, I saw the hands of a small child gripping onto the handrail of the upward escalator, but the child himself was out of view. I realized that he was riding up on the *outside* of the escala-

* When I reflect on the many times my friends have helped me out over the years, I see that nothing has meant more to me or stuck with me more than when their generosity was completely uncalculated. For example, I remember visiting my friend Sid in college and accidentally spending all my money before I'd found a way to get home. When I asked Sid (who was perennially broke) if he could help pay for a ticket, he jammed his hand in his pocket and handed over all the cash he had. He made no effort to count it, had no clue how much it was, and wasn't interested in finding out.

tor and had apparently been carried too far up and was now afraid to drop off. This was a major problem, as the top of the escalator was two stories high, and three-quarters of the way up there was a decorative pole abutting the escalator that would soon dislodge the boy from his rather tenuous hold.

It was an easy matter to jump over the handrail to the other escalator, so I ran down and reached the little boy a few yards before he got to the pole. He weighed almost nothing, and I easily picked him up by his arms from the far side of the escalator. But here's the stupid part: I risked gumming up the whole operation by virtue of the fact that I was on autopilot, in a mindlessly helpful state.

As I ran down the escalator, rather than chucking my ice cream to the ground below or dropping it on the escalator, I transferred it to my pinkie finger so I could continue to hold it while I reached over the handrail to grab the boy's arms. I'm sure that this process slowed me down a bit, and I'm also sure that it was harder to grab him while I was simultaneously trying to avoid dropping my ice cream. The ice cream also nearly got me punched in the face, as by now the boy's father had heard his wife screaming and saw me riding the escalator away from her, holding their small child and carrying an ice-cream cone.

I can only imagine what was running through his mind, but thankfully, he decided to withhold judgment and simply ran down the escalator and grabbed his child from me. When the rush of the event was over, I sat there looking at the ice-cream cone in my hand and wondering how I could have been so stupid. In retrospect, my effort to help the little boy was clearly conducted on autopilot. Without the cognitive wherewithal to weigh the costs and benefits of holding on to the ice-cream cone, I automatically followed the standard rules of politeness not to throw food on the ground. Indeed, cases of helping gone awry are so common that many countries (and U.S. states) have adopted "Good Samaritan

Laws," which protect would-be helpers from lawsuits if they make matters worse.*

The desire to cooperate is a powerful force, and the outcome of these evolutionary pressures is a motivational system that is highly attuned to helping others. We want the grandkids to come over for Sunday lunch, and we gain genuine satisfaction from helping family and friends. In a recent World Values Survey of nearly a hundred countries, people around the world rated family as the single most important thing in their lives, and this makes perfect evolutionary sense. But the altruistic satisfaction we gain from cooperating extends beyond kin and even close friends to our entire community.

As I discuss in chapter 3, our communities are much larger now than the ones in which we originally evolved, but the psychological principles that link us to our community have the same effects they've always had. In that sense, we haven't changed in any fundamental way from our hunter-gatherer selves. Integration with our community was and is one of the keys to living the good life. Unfortunately, as we've become wealthier and more reliant on technology, we have simultaneously reduced our reliance on one another and thereby unintentionally disrupted our integration with our neighbors and larger community.

Some people find it easy to enmesh themselves into new communities, and thus frequent moves don't impinge on their life satisfaction. Extroverts are such people, as they enjoy the company of others whom they don't know well, and view getting to know people as an opportunity. Introverts, in contrast, find meeting lots of new people difficult and unpleasant, so moving to a new neighborhood, city, or state has a major impact on their life satisfaction. Integration with one's community is important to everyone's life satisfaction, but how this is best achieved differs among people.

* For example, how would I have explained myself to a jury who wanted to know why I had let the child fall instead of the ice-cream cone?

For extroverts, pulling up stakes and heading across the country or around the world is of lesser consequence. For introverts, there will be a price to pay for frequent life moves, and career or schooling opportunities in new communities must be considered carefully in light of their offsetting costs.

It's important to keep in mind that "offsetting costs" are not just temporary feelings of unhappiness. As we discuss in the last chapter, happiness is the body's cue to allow the immune system to function at peak efficiency, and frequent unhappiness has long-term health costs. The importance of community integration for health can be seen in Figure 10.5, from research by Shigehiro Oishi and Ulrich Schimmack. This graph shows that frequent residential moves in childhood lead to accumulating damage to the health of introverts. Even though the damage occurs in childhood, the data suggest that they never fully recover. Introverts who moved frequently during childhood suffer greater late-life mortality, while extroverts are unaffected by childhood moves.

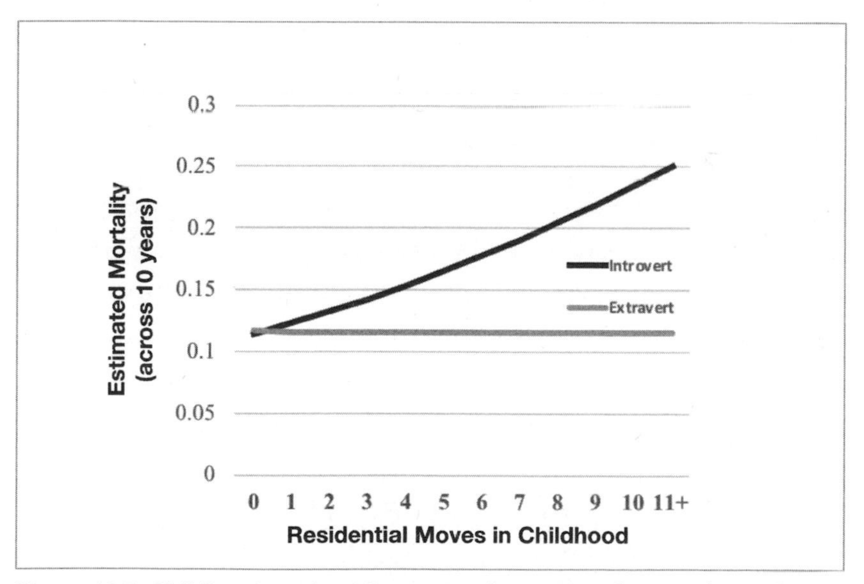

Figure 10.5. Childhood residential moves and mortality for ages 60–70 for introverts and extroverts. (Adapted from Oishi and Schimmack, 2010)

These data reflect the costs of breaking connections with our community, but of course most people are focused on creating and sustaining such connections. Life satisfaction is achieved by being embedded in your community and by supporting community members who are in need. This explains why the same people who adamantly oppose welfare are often incredibly generous in supporting downtrodden members of their own communities. When they are assured that the need is genuine and that they are not being taken for a ride, people of all political stripes endeavor to help their fellow humans.

Charity supports those in need, but feeling you have made a contribution to your community is the more important manifestation of this effect. We need to be of value to our community—our ancestors who were not of value were at risk of ostracism and hence death—and the most obvious way to be of value is to produce more than you cost. This calculus may not be conscious, but it drives us to be contributing members of society. When someone asks how you'd like to be remembered after you're gone, what they are really asking about is the nature of your contribution to your community.

Happiness and Learning

As I've mentioned throughout this book, humans actively learn most of what we need to know to survive and thrive. We're born knowing very little, with a brain that is really only half-baked but that would be too large to birth if we waited for it to be fully cooked. As a consequence, we have an inordinately long period of development before we can become viable, contributing members of our community. Unlike baby wildebeests, which can run from predators soon after they are born, we're pretty much anyone's potential snack for over a decade.

Our long period of development is consumed almost entirely by learning the means of survival used by our group. As discussed in chapter 2, our flexibility has enabled us to colonize the entire planet. But our flexibility also means that there is no way we can rely on inborn knowledge or instinct to survive. As a consequence, evolution has ensured that learning is tightly linked to our motivational system; humans all over the world love to learn. Curiosity is one of our fundamental drives, and the satisfaction associated with learning and mastering something new is universal.

The motivational importance of curiosity is widely understood, but there are two important forms of learning (and therefore two important sources of life satisfaction) that people often fail to recognize: play and storytelling. Play is universal in humans just as it is among our mammalian cousins. Play is most common in animals before they reach maturity, because it is a form of learning the rules and strategies of adult life. Play also helps humans and other animals learn to cooperate, as their early interactions with one another are positive and teach them who reciprocates and who does not. Play provides opportunities for young males to learn to compete for females, it teaches strategies for eventually rising in the adult hierarchy, and it helps juveniles learn to hunt prey and escape predators. Kittens like nothing better than to pounce on one another in mock battles; little boys enjoy this as well; and all human children engage in endless hours of pretend play, sport, and games. Because humans are unique in the extraordinary amount of information they must learn, the importance of play has extended beyond our childhood and into our adult lives. In the absence of play, life is a lot less fun.

The importance of play is common to all mammals (and the occasional bird and reptile), but storytelling is unique to humans. A major advantage of human learning is that our incredible communicative abilities enable us to incorporate the learning and accomplishments of others into our understanding of the world. The oldest and most important form of this type of learning is our rich oral tradition. Storytelling can be found in all human cultures, and surely began when our hunter-gatherer ancestors sat around the fire at the end of the day and regaled each other with tales of their experiences. Storytelling around the fire is still an important activity in hunter-gatherer communities; it's a time when the immediate economic and social concerns of the day are put aside and people focus on the broader patterns, rules, and lessons for achieving a productive life in their community.

Good storytellers gained status in their group because they were valued for the entertainment and learning they provided. In our world these roles are occupied by comedians, talk-show hosts, filmmakers, ministers, imams, rabbis, authors, academics, and political pundits who entertain, inform, and transmit social norms to their listeners. Everyone evolved a proclivity to enjoy listening to stories, as hearing about the trials and tribulations of others provides a risk-free way to learn the difficult and sometimes expensive lessons garnered from their experiences. Finally, storytelling connected community members to each other through shared emotional experiences, a sense of shared reality, and a common knowledge of how to approach the world.

When I tell my story about accidentally peeing in the sink at a Springsteen concert, and listeners share my embarrassment and laughter, I feel validated, and my connection to my social group is a little bit stronger. When I listen to the wonderful, outrageous, or scary stories of my friends and discover how they turned out, I learn what to do if I ever find myself in similar circumstances. For these reasons, listening to and telling stories are two important sources of happiness and life satisfaction.

Happiness, Personality, and Development

As I suggest earlier in this chapter, there is more than one way to be a successful human, and hence more than one route to happiness. If I'm big and strong, I might attract a mate through sports or other physical competitions, but if I'm small and weak, I'd be better off using my humor or kindness. Such differences in brawn, brains, and personality can also make me suited to different types of careers. People tend to choose a strategy of best fit by pursuing activities that rely on their strengths and avoid their weaknesses. Because our motivational system is attuned to our potential for success and the degree to which our group values us, different activities make different people happy. If I'm better at art than athletics, I'll probably find greater satisfaction from painting than from playing football.

These individual differences, in turn, also fluctuate over the course of our life span. When we are children, it's very difficult to contribute to our group, and hence children gain most of their happiness from their activities with one another and from acceptance by their parents and peers. Because their unique features are more likely to be noticed by others than their common ones, their sense of self becomes tied to the ways they can differentiate themselves positively in their group.

Even in childhood, people start to focus on and develop the unique talents that will make them most productive when they reach adulthood, but of course their childlike worldview will often take them in odd directions. I remember when my daughter learned to read by listening in while I was helping her older brother with his homework, and I tried to get her to brag about it to her preschool teacher. "What did you learn to do yesterday?" I asked her when we sat down with her teacher. She looked puzzled for a moment, and then brightened up and announced, "I learned to slide down the fire pole!" As this example reveals, positive distinctiveness is not always achieved by children in a manner that reflects adult priorities.

Once we reach adulthood, our contribution to the community becomes important, and we start to rely on the skills we've learned in an effort to rise in the hierarchy and be valued by members of our group. These skills are relatively stable in most people throughout young and middle adulthood, but they begin to show notable changes in late life. At that point, physical skills typically decline. Offsetting their deteriorating physical skills, older adults tend to have a larger body of knowledge, which made them valuable in our ancestral past.

Unfortunately, in our modern and rapidly changing world, the knowledge of older adults can become obsolete. But the evolutionary pressures that push us to contribute to our group don't change just because we have gotten older, and many older adults strive to find ways to make a positive impact on their community. Issues of legacy and care for others become increasingly important among older adults, as these are often their greatest opportunities to remain connected and helpful to their community. In our quest for the good life, we should remember that sitting back and putting our feet up may sound great, but retirement is likely to be more satisfying if we can also find a way to make a contribution to others. When we're overwhelmed by work and other commitments and have no time for ourselves, a permanent vacation sounds like a great idea. But don't be fooled by retirement brochures that promote only leisure activities—for most of us, that's not as good as it sounds.

Pitfalls of a Modern World

Pondering the good life may be a uniquely modern pastime, but the ways we achieve it are by following the ancient strategies that made our ancestors successful. Sex and food, parenthood and play, mastery and storytelling, friendship and kin, hearth and home, community and contribution—these were the keys to success in our past and they remain the keys to our happiness today. Nonetheless, our modern world provides many new opportunities for happiness, and it is not always clear when the modern version is just as good as the original.

For example, movies and television have replaced many aspects of storytelling, and both are a lot of fun. But storytelling is much more than just relating a series of events, and movies and television don't connect people to each other in the same way that conversations do (unless we talk about them later, of course). To some degree, movies and television are thus a phenotypic indulgence (like potato chips), so it is no surprise that people rarely cite television programs as an important source of life satisfaction, no matter how much they love their favorite show. Anecdotal evidence suggests that books might be more likely than TV to have an important and lasting effect on us. If that is true, a critical part of storytelling may be the imaginative and generative processes taking place in our minds when someone else tells a tale that we cannot directly experience. But even books are typically less memorable and important than the stories we tell and are told, because reading is typically done alone.

Other aspects of our modern world that mimic important ancestral experiences provide much thinner gruel and leave us a lot less satisfied. Drugs and alcohol are probably the most notable examples of phenotypic indulgences, as they go straight to the brain regions responsible for pleasure without providing the physical or experiential basis from which that pleasure was meant to emanate. After drugs and alcohol, junk food comes in a close second, as the sugar, fat, and salt that our ancestors desperately sought in the past are overabundant today. Sadly, our struggle is now against what was once a healthy goal: to eat as much sugar, fat, and salt as possible.

The price we pay when we have too much of a good thing brings us to the final lesson I want to highlight from our evolutionary history. Long-term sexual relationships were our ancestors' best recipe for raising successful offspring, and as a result we find long-term relationships particularly rewarding. When we partner with the right person, we have our best shot at a lasting increase in happiness. But evolution also gave us a preference for novel partners, as both men and women gain reproductive benefits when they put their genetic eggs into more than one basket.

The problem is that novel partners were relatively rare in our ancestral environment, as we spent our entire lives in the same small group of people. But we now live in a world in which novel partners—like fat, salt, and sugar—are in unending supply and serve as a constant temptation for us to abandon our current relationship to try out a new and more exciting one. Of course, the new relationship will soon become old, as the allure of novelty is fleeting by definition and thus eventually unsatisfying. Rather remarkably, that obvious fact doesn't prevent people from being serial monogamists now, just as it didn't prevent our ancestors from adopting a similar mating strategy back in the day.

Most people are better at avoiding temptation than resisting it, and, sure enough, people who escape the lure of novelty usually achieve this goal by not exposing themselves to it. Marriages last longer in rural areas than in cities, and much longer still if you're a nobody than if you're a famous actor or rock star. And these findings bring us back to the German folk saying about the inevitable disappointment we experience when we achieve our goals. Universal adoration and fame are some of the most common dreams of people all over the world, but you need only reflect on the turbulent lives and repeated divorces of celebrities to realize how much happier you are being unknown.

Getting to the Good Life in "Ten Easy Steps"

If anyone who doesn't know you tells you there are ten easy steps to happiness, you're being taken for a ride. As I've just discussed, not only is there no such thing as lasting happiness, but paths to happiness differ for different people. Nevertheless, approaching happiness from an evolutionary perspective can help us achieve it, at least some of the time, and can also help us understand happiness itself. So, I'm not really offering ten easy steps here, but in service of these goals, I summarize the lessons of the last two chapters in ten essential points.

1. **Stay present.** Our proclivity to live in the future disrupts our ability to enjoy the present, particularly when the present provides unexpected pleasures. If you fail to find happiness in the pleasures of daily life, then a program of mindfulness or meditation might help you. Learning to live in the present might also reduce your stress levels if you tend to worry about the future. Keep in mind, however, that living in the now is a lot more difficult than it sounds, largely because you are trying to shut down one of the most important skills that evolution gave you in your capacity to plan for the future.

2. **Seek out sweet moments.** It's almost impossible to become *permanently* happier, but that doesn't mean you can't have more fun in your life. Achieving our evolutionary imperatives brings us positive emotions that range from contentment to great joy. We just need to be prepared for the fact that these feelings won't last. But who wouldn't opt for more frequent happy moments?

3. **Guard your happiness to stay healthy.** Happiness is critical for physical health. If you are sacrificing your happiness for something that isn't incredibly important, you should ask yourself how long this situation has lasted and how long it's likely to last in the future. Short-term sacrifices can be sensible, but long-term sacrifices should be avoided if at all possible. If you must sacrifice your happiness to achieve other goals, try to prepare a time line and stick to it. Otherwise, you might wake up one day to find your short-term sacrifice has been going on for years, and your happiness and health are a thing of the past.

4. **Accumulate experiences, not stuff.** The great times you have become a part of you; the great things you own gather dust or become trash. That said, if your memory for your lived experiences is as bad as mine, you might struggle to recall many of the great times you've had, and then find yourself at risk of owning nothing and remembering little. A simple solution to this problem is to take pictures, and even purchase the occasional doodad from your travels or adventures. By scattering these reminders around your home or office, you can relive your great times and laugh at the adventures that went awry.

5. **Prioritize food, friends, and sexual relationships.** These three things are the basis of day-to-day happiness. Note that there is no mention here of money or freedom. There's nothing wrong with having lots of cash and autonomy, but their pursuit shouldn't interfere with opportunities to enjoy good food, sex, and friends. These three things are most likely to provide the happy experiences that accumulate into a life worth living.

6. **Cooperate.** Working together with family, friends, and colleagues to achieve mutual goals is one of the most important sources of life satisfaction. Your achievements won't make you permanently happier, but cooperation is inherently rewarding and provides a foundation for life satisfaction. Happiness doesn't emerge only from leisure and fun, but also from work and productivity, particularly when you are satisfying your evolutionary imperative of cooperating with others. Not all the work we do is meaningful, as life has necessary drudgery, but working with people you trust and admire lightens the load.

7. **Embed yourself in community.** Give careful thought to any decisions that require you to pull up roots and go somewhere else. We evolved to be curious, so new people and new places

are forever enticing. But you don't need to abandon old friends to meet new people and see new places. Even if you have a strong wanderlust, you should try to retain your connections to your community.

8. **Learn new things.** Learning is a lifelong source of happiness, and play and storytelling are two important sources of learning. At all stages of life, from childhood to young adulthood through to midlife and old age, we enjoy mastering new things. If you choose your activities thoughtfully, you can enjoy the process of learning up to your last healthy days on this earth.

9. **Play to your strengths.** There are many roads to happiness, but almost all of them are found by pursuing your particular strengths, which are likely to change over time. Change is intimidating for almost everyone, as it requires us to move from the known to the unknown, and hence from the predictable to the unpredictable. For this reason, many people remain too long with jobs or hobbies that once suited them but do not anymore. Just because you once loved something doesn't mean you are destined to always feel that way. Your changing sources of happiness are probably telling you that your old life doesn't suit you anymore.

10. **Seek the original source.** Our modern world provides numerous opportunities for happiness that resemble but do not duplicate the original sources. Some are perfectly fine (e.g., TV and movies), some probably do more harm than good (e.g., alcohol, drugs, and junk food), but none is as good as the ancestral originals. Time with family and friends sits at the top of our species' checklist and is our best recipe for happiness.

Epilogue

Evolution is not a warm and fuzzy concept. Whoever leaves behind the most offspring wins, no matter how that goal is achieved. So, it's no surprise that the process itself is often brutal. I remember watching a nature show on TV in which a pack of hyenas tore a baby zebra limb from limb without even bothering to kill it first. I felt sick to my stomach, but for the hyenas, it was just an afternoon snack, and I'm sure they gave it no thought once they finished the last tidbit. Animals that arrive at effective solutions to their problems pass on their genes, and with them, their solutions—whether they're vicious killers like those hyenas or adorable vegetarians like the baby zebra they consumed. Indeed, vicious and adorable, good and bad, moral and immoral—all these are human constructs that don't exist in the natural world. Evolution is amoral.

Why am I reminding you of this? Because the evolutionary pressures inherent in such a world could easily have brought us to a miserable place. Our chimp cousins rarely look out for one another the way we do, and the same holds true for our baboon second cousins. If you read Robert Sapolsky's wonderful book on savannah baboons, *A Primate's Memoir*, you'll see that the life of a baboon isn't much fun unless you're the alpha, with everyone endlessly harassing the monkey below them in the hierarchy. Their solution to the challenges of life on the savannah could just as easily have been our own, but as luck would have it, *Australopithecines* evolved to protect themselves by working together. *Homo erectus* then expanded on their ancestors' loose-knit cooperation with division of labor, and the resultant interdependence gave us a life strategy that was not only effective but kind as well.

One of the most disconcerting aspects of evolution is the enormous role played by random chance. Our existence as a species is the result of innumerable rolls of the dice, every one of which had to go our way. The most trivial perturbations in our past would have changed everything. If our parents had felt amorous on a different night, or if other sperm happened to win the race to fertilize our mothers' eggs, I wouldn't be writing this and you wouldn't be reading it. The probability that either of us got a chance to live at all is vanishingly small, and yet here we are. As Richard Dawkins puts it in his fascinating book *Unweaving the Rainbow*, "We are going to die, and that makes us the lucky ones."

But the mere fact that we get to live is not what makes us lucky. Many animals live a life that I would just as soon forgo, not because it ends in tragedy, as it did for that baby zebra, but because their approach to living is one of endless conflict. Imagine being a seagull and spending your entire life fighting other seagulls for scraps. What makes us so lucky is the pure happenstance that we evolved to be (mostly) good to one another.

Our cooperative nature also set the stage for the evolution of our amazing brain. Our sociality made us smarter individually, but, far more important, it connected our minds to others' minds in a manner that massively increased our knowledge and computing power. As a result, we long ago surpassed the predators that hunted us on the savannah, and are now holding most of the pathogens at bay that are a much greater threat than predators ever were. For the first time in history, we no longer bury almost half our children before they reach adulthood. Evolution is brutal, but those of us with the good fortune to live in established democracies have used the tools that evolution gave us to create unprecedentedly safe and satisfying lives. We evolved a psychology that continually searches for something better, but a moment's reflection reveals that it's hard to ask for much more than that.

Acknowledgments

The cooperativeness we evolved on the savannah not only brought us to the top of the food chain, but also enabled the scientific enterprise. Like every other human on this planet, I'm a product of many teachers, mentors, and collaborators, and this book represents a massive cooperative enterprise. There is no way I could have written it on my own, and I didn't even try. First of all, huge thanks are owed to Lauren Sharp, my agent at Aevitas Creative Management, who contacted me based on a fifteen-minute podcast at *Harvard Business Review* and encouraged me to write the book in the first place. Lauren also helped me develop the project. Another huge thanks to my editors at Harper Wave: Hannah Robinson, who provided superb editorial guidance; Jenna Dolan, who rescued my grammar and punctuation more times than I care to admit; and Sarah Murphy, who worked on part of the book before moving to another position. And, critically, thanks to Karen Rinaldi at Harper Wave, who decided to invest in me and this book.

My agent and editors played a central role, but so did my friends and family, who suffered through numerous preliminary drafts that were too embarrassing to show Lauren and Hannah. To begin with she who suffered most (and not just because I'd get in trouble otherwise), Courtney read the first draft of every chapter, pointing out when it was boring or unclear, but most notably trying to help me be a little chattier and a little less academic (aka dry and stuffy). After Courtney read them, the chapters went out to a long list of friends and family, and I owe them a huge thank-you: Roy Baumeister, Rob Brooks, Adam Bulley, Steve Fein, Mickey Inzlicht, Pamela Krones, Matt Lieberman, Dave Marshall, Elizabeth Marx, Glen McBride, Amanda Niehaus, Sam Pearson, Tiko Shah, Thomas Suddendorf, and Meris Van de Grift; Arndt, Cathy, Frank, Karin, Marianne, Maya, Paul, and Ted vH; and Henry Wellman, Robbie Wilson, Matti Wilks, and Brendan Zietsch. This book is much better because of them.

I am very fortunate to be part of a superb group of scholars at the University of Queensland, most notably in the Centre for Psychology and Evolution. The ideas on which this book is based have largely been formed in discussions, presentations, and debates in the Centre, and I'm very grateful to all the members and visitors over the past decade, particularly Thomas Suddendorf and Brendan Zietsch. Finally, I owe a big debt to my collaborators on the academic work that forms the basis of most of this book (see the papers listed in the reference section). Without them, the book wouldn't exist.

References

PROLOGUE

Boesch, C. "Cooperative Hunting in Wild Chimpanzees." *Animal Behaviour* 48 (1994): 653–67.

Copeland, S. R., M. Sponheimer, D. J. de Ruiter, J. A. Lee-Thorp, D. Codron, P. J. le Roux . . . and M. P. Richards. "Strontium Isotope Evidence for Landscape Use by Early Hominins." *Nature* 474 (2011): 76–78.

Hill, K. R., R. S. Walker, M. Božičević, J. Eder, T. Headland, B. Hewlett . . . and B. Wood. "Co-Residence Patterns in Hunter-Gatherer Societies Show Unique Human Social Structure." *Science* 331 (2011): 1286–89.

Kittler, R., M. Kayser, and M. Stoneking. "Molecular Evolution of *Pediculus humanus* and the Origin of Clothing." *Current Biology* 13 (2003): 1414–17.

Lahdenperä, M., V. Lummaa, S. Helle, M. Tremblay, and A. F. Russell. "Fitness Benefits of Prolonged Post-Reproductive Lifespan in Women." *Nature* 428 (2004): 178–81.

Nesse, R. M., and G. C. Williams. *Why We Get Sick: The New Science of Darwinian Medicine.* New York: Vintage, 1995.

Thornton, A., and K. McAuliffe. "Teaching in Wild Meerkats." *Science* 313 (2006): 227–29.

Tomasello, M. *A Natural History of Human Morality.* Cambridge, MA: Harvard University Press, 2016.

von Hippel, W., and D. M. Buss. "Do Ideologically Driven Scientific Agendas Impede the Understanding and Acceptance of Evolutionary Principles in Social Psychology?" In Lee Jussim and Jarret T. Crawford, eds. *The Politics of Social Psychology.* Frontiers in Psychology series. New York: Routledge, 2017.

Yang, F., Y.-J. Choi, A. Misch, X. Yang, and Y. Dunham. "In Defense of the Commons: Young Children Negatively Evaluate and Sanction Free-Riders." *Psychological Science* (July 2018), https://doi.org/10.1177%2F0956797618779061.

1: EXPELLED FROM EDEN

Ashton, B. J., A. R. Ridley, E. K. Edwards, and A. Thornton. "Cognitive Performance Is Linked to Group Size and Affects Fitness in Australian Magpies." *Nature* 554 (2018): 364–67.

Bates, L. A., K. N. Sayialel, N. W. Njiraini, J. H. Poole, C. J. Moss, and R. W. Byrne. "African Elephants Have Expectations About the Locations of Out-of-Sight Family Members." *Biology Letters* 4 (2008): 34–36.

Bingham, P. M. "Human Evolution and Human History: A Complete Theory." *Evolutionary Anthropology* 9 (2000): 248–57.

Boesch, C. "Cooperative Hunting in Wild Chimpanzees." *Animal Behaviour* 48 (1994): 653–67.

———. "The Effects of Leopard Predation on Grouping Patterns in Forest Chimpanzees." *Behaviour* 117 (1991): 220–42.

Calvin, W. H. "Did Throwing Stones Shape Hominid Brain Evolution?" *Ethology and Sociobiology* 3 (1982): 115–24.

Coppens, Y. "East Side Story: The Origin of Humankind." *Scientific American* 270 (1994): 88–95.

Crompton, R. H., T. C. Pataky, R. Savage, K. D'Août, M. R. Bennett, M. H. Day . . . and W. I. Sellers. "Human-Like External Function of the Foot, and Fully Upright Gait, Confirmed in the 3.66 Million Year Old Laetoli Hominin Footprints by Topographic Statistics, Experimental Footprint-Formation and Computer Simulation." *Journal of the Royal Society Interface* 9 (2012): 707–19.

Dart, R. A. "*Australopithecus africanus*: The Man-Ape of South Africa." *Nature* 115 (1925): 195–99.

Dunbar, R. I., and S. Shultz. "Evolution in the Social Brain." *Science* 317 (2007): 1344–47.

Frith, U., and C. Frith. "The Social Brain: Allowing Humans to Boldly Go Where No Other Species Has Been." *Philosophical Transactions of the Royal Society B* 365 (2010): 165–75.

Gilby, I. C. "Meat Sharing Among the Gombe Chimpanzees: Harassment and Reciprocal Exchange." *Animal Behaviour* 71 (2006): 953–63.

Hare, B., and M. Tomasello. "Chimpanzees Are More Skillful in Competitive than in Cooperative Cognitive Tasks." *Animal Behaviour* 68 (2004): 571–81.

Hart, D., and R. W. Sussman. *Man the Hunted: Primates, Predators, and Human Evolution*. Boulder, CO: Westview Press, 2005.

Humphrey, N. "The Social Function of Intellect." In P. P. G. Bateson and R. A. Hinde, eds. *Growing Points in Ethology*. Cambridge, UK: Cambridge University Press, 1976, pp. 303–13.

Isaac, B. "Throwing and Human Evolution." *African Archaeological Review* 5 (1987): 3–17.

Kaiho, K., and N. Oshima. "Site of Asteroid Impact Changed the History of Life on Earth: the Low Probability of Mass Extinction." *Scientific Reports* 7 (2017): 148–55.

Kortlandt, A. *New Perspectives on Ape and Human Evolution.* Amsterdam: Stichting voor Psychobiologie, 1972.

Lieberman, M. D. *Social: Why Our Brains Are Wired to Connect.* New York: Oxford University Press, 2013.

Marzke, M. W. "Joint Functions and Grips of the *Australopithecus afarensis* Hand, with Special Reference to the Region of the Capitate." *Journal of Human Evolution* 12 (1983): 197–211.

Pinker, S. "The Cognitive Niche: Coevolution of Intelligence, Sociality, and Language." *Proceedings of the National Academy of Sciences* 107 (2010): 8993–99.

Powell, L. E., K. Isler, and R. A. Barton. "Re-Evaluating the Link Between Brain Size and Behavioural Ecology in Primates." *Proceedings of the Royal Society B* 284 (2017): 1765.

Pruetz, J. D., and P. Bertolani. "Savanna Chimpanzees, *Pan troglodytes verus*, Hunt with Tools." *Current Biology* 17 (2007): 412–17.

Pruetz, J. D., and S. Lindshield. "Plant-Food and Tool Transfer Among Savanna Chimpanzees at Fongoli, Senegal." *Primates* 53 (2012): 133–45.

Roach, N. T., M. Venkadesan, M. J. Rainbow, and D. E. Lieberman. "Elastic Energy Storage in the Shoulder and the Evolution of High-Speed Throwing in *Homo*." *Nature* 498 (2013): 483–87.

Tomasello, M. *A Natural History of Human Morality.* Cambridge, MA: Harvard University Press, 2016.

Whiten, A., and R. W. Byrne. "Tactical Deception in Primates." *Behavioral and Brain Sciences* 11 (1988): 233–73.

Williams, K. D. *Ostracism: The Power of Silence.* New York: Guilford Press, 2002.

Young, R. W. "Evolution of the Human Hand: The Role of Throwing and Clubbing." *Journal of Anatomy* 202 (2003): 165–74.

2: OUT OF AFRICA

Baumeister, R. F. *The Cultural Animal: Human Nature, Meaning, and Social Life.* New York: Oxford University Press, 2005.

Berna, F., P. Goldberg, L. K. Horwitz, J. Brink, S. Holt, M. Bamford, and M. Chazan. "Microstratigraphic Evidence of In Situ Fire in the Acheulian Strata of Wonderwerk Cave, Northern Cape Province, South Africa." *Proceedings of the National Academy of Sciences* 109 (2012): E1215–E1220.

Boesch, C. "Teaching Among Wild Chimpanzees." *Animal Behaviour* 41 (1991): 530–32.

Boyd, R., and P. J. Richerson. "Culture and the Evolution of the Human Social Instincts." In S. Levinson and N. Enfield, eds. *Roots of Human Sociality.* Oxford, England: Berg, 2006, pp. 453–77.

Diez-Martín, F., P. S. Yustos, D. Uribelarrea, E. Baquedano, D. F. Mark, A. Mabulla, and J. Yravedra. "The Origin of the Acheulean: The 1.7 Million-Year-Old Site of FLK West, Olduvai Gorge (Tanzania)." *Scientific Reports* 5 (2015): 17839.

Ding, X. P., H. M. Wellman, Y. Wang, G. Fu, and K. Lee. "Theory-of-Mind Training Causes Honest Young Children to Lie." *Psychological Science* 26 (2015): 1812–21.

Domínguez-Rodrigo, M. "Hunting and Scavenging by Early Humans: The State of the Debate." *Journal of World Prehistory* 16 (2002): 1–54.

Fiddes, I. T., G. A. Lodewijk, M. Mooring, S. R. Salama, F. M. J. Jacobs, and D. Haussler. "Human-Specific NOTCH2NL Genes Affect Notch Signaling and Cortical Neurogenesis." *Cell* 173 (2018): 1356–69.

Gallotti, R., and M. Mussi. "The Unknown Oldowan: ~1.7-Million-Year-Old Standardized Obsidian Small Tools from Garba IV, Melka Kunture, Ethiopia." *PLoS One* 10, no. 12 (2015): e0145101. https://doi.org/10.1371/journal.pone.0145101.

Gibbons, A. "Why Humans Are the High-Energy Apes." *Science* 352 (2016): 639.

Goren-Inbar, N., A. Lister, E. Werker, and M. Chech. "A Butchered Elephant Skull and Associated Artifacts from the Acheulian Site of Gesher Benot Ya'aqov, Israel." *Paléorient* (1994): 99–112.

Harcourt, A. H. "Human Phylogeography and Diversity." *Proceedings of the National Academy of Sciences* 113 (2016): 8072–78.

Harmand, S., J. E. Lewis, C. S. Feibel, C. J. Lepre, S. Prat, A. Lenoble . . . and H. Roche. "3.3-million-year-old Stone Tools from Lomekwi 3, West Turkana, Kenya." *Nature* 521 (2015): 310–15.

Henrich, J. *The Secret of Our Success: How Culture Is Driving Human Evolution, Domesticating Our Species, and Making Us Smarter.* Princeton, NJ: Princeton University Press, 2015.

Horner, V., and A. Whiten. "Causal Knowledge and Imitation/Emulation Switching in Chimpanzees (*Pan troglodytes*) and Children (*Homo sapiens*)." *Animal Cognition* 8 (2005): 164–81.

Krupenye, C., F. Kano, S. Hirata, J. Call, and M. Tomasello. "Great Apes Anticipate That Other Individuals Will Act According to False Beliefs." *Science* 354 (2016): 110–14.

Nadel, D., A. Danin, R. C. Power, A. M. Rosen, F. Bocquentin, A. Tsatskin . . . and O. Barzilai. "Earliest Floral Grave Lining from 13,700–11,700-y-old Natufian Burials at Raqefet Cave, Mt. Carmel, Israel." *Proceedings of the National Academy of Sciences* 110 (2013): 11774–78.

Pinker, S. *The Better Angels of Our Nature: The Decline of Violence in History and Its Causes*. London: Penguin UK, 2011.

Roberts, W. A. "Are Animals Stuck in Time?" *Psychological Bulletin* 128 (2002): 473–89.

Roche, H., A. Delagnes, J. P. Brugal, C. Feibel, M. Kibunjia, V. Mourre, and P. J. Texier. "Early Hominid Stone Tool Production and Technical Skill 2.34 MYR Ago in West Turkana, Kenya." *Nature* 399 (1999): 57–60.

Shipton, C., and M. Nielsen. "Before Cumulative Culture." *Human Nature* 26 (2015): 331–45.

Stout, D., E. Hecht, N. Khreisheh, B. Bradley, and T. Chaminade. "Cognitive Demands of Lower Paleolithic Toolmaking." *PLoS One* 10, no. 4 (2015): e0121804.

Suddendorf, T. *The Gap: The Science of What Separates Us from Other Animals*. New York: Basic Books, 2013.

Suddendorf, T., and M. C. Corballis. "The Evolution of Foresight: What Is Mental Time Travel, and Is It Unique to Humans?" *Behavioral and Brain Sciences* 30 (2007): 299–313.

Suzuki, I., D. Gacquer, R. Van Heurck, D. Kumar, M. Wojno, A. Bilheu, A. Herpoel, et al. "Human-Specific NOTCH2NL Genes Expand Cortical Neurogenesis Through Delta/Notch Regulation." *Cell* 173 (2018): 1370–84.

Sznycer, D., J. Tooby, L. Cosmides, R. Porat, S. Shalvi, and E. Halperin. "Shame Closely Tracks the Threat of Devaluation by Others, Even Across Cultures." *Proceedings of the National Academy of Sciences* (2015): 14699.

Trivers, R. L. "The Evolution of Reciprocal Altruism." *Quarterly Review of Biology* 46 (1971): 35–57.

Wheeler, B. C. "Monkeys Crying Wolf? Tufted Capuchin Monkeys Use Anti-Predator Calls to Usurp Resources from Conspecifics." *Proceedings of the Royal Society of London B: Biological Sciences* 276 (2009): 3013–18.

Wiessner, P. W. "Embers of Society: Firelight Talk Among the Ju/'Hoansi Bushmen." *Proceedings of the National Academy of Sciences* 111 (2014): 14027–35.

Wrangham, R. *Catching Fire: How Cooking Made Us Human*. New York: Basic Books, 2009.

Wrangham, R., and R. Carmody. "Human Adaptation to the Control of Fire." *Evolutionary Anthropology: Issues, News, and Reviews* 19 (2010): 187–99.

If you are interested in further reading, chapters 1 and 2 are based on the following academic paper:

von Hippel, W., F. A. von Hippel, and T. Suddendorf. "Evolutionary Foundations of Social Psychology." In P. Van Lange, E. T. Higgins, and A. Kruglanski, eds. *Social Psychology: The Handbook of Basic Principles* (in press).

3: CROPS, CITIES, AND KINGS

Acemoglu, D., and J. A. Robinson. *Why Nations Fail: The Origins of Power, Prosperity, and Poverty*. New York: Crown Business, 2013.

Alesina, A., P. Giuliano, and N. Nunn. "On the Origins of Gender Roles: Women and the Plough." *Quarterly Journal of Economics* 128 (2013): 469–530.

Ambady, N., F. J. Bernieri, and J. A. Richeson. "Toward a Histology of Social Behavior: Judgmental Accuracy from Thin Slices of the Behavioral Stream." *Advances in Experimental Social Psychology* 32 (2000): 201–71.

Boehm, C. *Hierarchy in the Forest: The Evolution of Egalitarian Behavior.* Cambridge, MA: Harvard University Press, 2009.

Bollongino, R., O. Nehlich, M. P. Richards, J. Orschiedt, M. G. Thomas, C. Sell . . . and J. Burger. "2000 Years of Parallel Societies in Stone Age Central Europe." *Science* 342 (2013): 479–81.

Gibbons, A. "How Sweet It Is: Genes Show How Bacteria Colonized Human Teeth." *Science* 339 (2013): 896–97.

Lawler, A. "Uncovering Civilization's Roots." *Science* 335 (2012): 790–93.

Lippi, M. M., B. Foggi, B. Aranguren, A. Ronchitelli, and A. Revedin. "Multistep Food Plant Processing at Grotta Paglicci (Southern Italy) Around 32,600 cal BP." *Proceedings of the National Academy of Sciences* 112 (2015): 12075–80.

Liu, L., S. Bestel, J. Shi, Y. Song, and X. Chen. "Paleolithic Human Exploitation of Plant Foods During the Last Glacial Maximum in North China." *Proceedings of the National Academy of Sciences* 110 (2013): 5380–85.

Martins, Y., G. Preti, C. R. Crabtree, T. Runyan, A. A. Vainius, and C. J. Wysocki. "Preference for Human Body Odors Is Influenced by Gender and Sexual Orientation." *Psychological Science* 16 (2005): 694–701.

Mattison, S. M., E. A. Smith, M. K. Shenk, and E. E. Cochrane. "The Evolution of Inequality." *Evolutionary Anthropology: Issues, News, and Reviews* 25 (2016): 184–99.

Nisbett, R. E., and D. Cohen. *Culture of Honor: The Psychology of Violence in the South.* Boulder, CO: Westview Press, 1996.

Pringle, H. "The Ancient Roots of the 1%." *Science* 344 (2014): 822–25.

Starmans, C., M. Sheskin, and P. Bloom. "Why People Prefer Unequal Societies." *Nature Human Behaviour* 1 (2017): 1–7.

Willcox, G. "The Roots of Cultivation in Southwestern Asia." *Science* 341 (2013): 39–40.

Winterhalder, B. "Work, Resources and Population in Foraging Societies." *Man* (1993): 321–40.

Zerjal, T., Y. Xue, G. Bertorelle, R. S. Wells, W. Bao, S. Zhu . . . and P. Li. "The Genetic Legacy of the Mongols." *American Journal of Human Genetics* 72 (2003): 717–21.

4: SEXUAL SELECTION AND SOCIAL COMPARISON

Bräuer, J., and D. Hanus. "Fairness in Non-human Primates?" *Social Justice Research* 25 (2012): 256–76.

Brosnan, S. F., and F. B. De Waal. "Monkeys Reject Unequal Pay." *Nature* 425 (2003): 297–99.

Brown, C., M. P. Garwood, and J. E. Williamson. "It Pays to Cheat: Tactical Deception in a Cephalopod Social Signaling System." *Biology Letters* (2012): rsbl20120435.

Buss, D. M., and D. P. Schmitt. "Sexual Strategies Theory: An Evolutionary Perspective on Human Mating." *Psychological Review* 100 (1993): 204.

Darwin, C. *The Descent of Man, and Selection in Relation to Sex.* 2nd ed. London: John Murray, 1874.

Engelmann, J. M., J. B. Clift, E. Herrmann, and M. Tomasello. "Social Disappointment Explains Chimpanzees' Behaviour in the Inequity Aversion Task." *Proceedings of the Royal Society B* 284 (2017): 20171502.

Galperin, A., M. G. Haselton, D. A. Frederick, J. Poore, W. von Hippel, D. M. Buss, and G. C. Gonzaga. "Sexual Regret: Evidence for Evolved Sex Differences." *Archives of Sexual Behavior* 42 (2013): 1145–61.

Kuziemko, I., R. W. Buell, T. Reich, and M. I. Norton. "'Last-Place Aversion': Evidence and Redistributive Implications." *Quarterly Journal of Economics* 129 (2014): 105–49.

Leimgruber, K. L., A. G. Rosati, and L. R. Santos. "Capuchin Monkeys Punish Those Who Have More." *Evolution and Human Behavior* 37 (2016): 236–44.

Tesser, A. "Toward a Self-Evaluation Maintenance Model of Social Behavior." *Advances in Experimental Social Psychology* 21 (1988): 181–27.

Trivers, R. "Parental Investment and Sexual Selection." In B. Campbell, ed. *Sexual Selection and the Descent of Man.* New York: Aldine de Gruyter, 1972.

von Rueden, C., M. Gurven, and H. Kaplan. "Why Do Men Seek Status? Fitness Payoffs to Dominance and Prestige." *Proceedings of the Royal Society of London B: Biological Sciences* 278 (2010): 2223–32.

Zahavi, A. "Mate Selection: A Selection for a Handicap." *Journal of Theoretical Biology* 53 (1975): 205–14.

5: *HOMO SOCIALIS*

Anderson, C., S. Brion, D. A. Moore, and J. A. Kennedy. "A Status-Enhancement Account of Overconfidence." *Journal of Personality and Social Psychology* 103 (2012): 718–35.

Boehm, C. *Moral Origins: The Evolution of Virtue, Altruism, and Shame.* New York: Soft Skull Press, 2012.

Boysen, S. T., and G. G. Berntson. "Responses to Quantity: Perceptual Versus Cognitive Mechanisms in Chimpanzees (*Pan troglodytes*)." *Journal of Experimental Psychology: Animal Behavior Processes* 21 (1995): 82.

Brady, W. J., J. A. Wills, J. T. Jost, J. A. Tucker, and J. J. Van Bavel. "Emotion Shapes the Diffusion of Moralized Content in Social Networks." *Proceedings of the National Academy of Sciences* 114 (2017): 7313–18.

Ditto, P. H., and D. F. Lopez. "Motivated Skepticism: Use of Differential Decision Criteria for Preferred and Nonpreferred Conclusions." *Journal of Personality and Social Psychology* 63 (1992): 568–84.

Epley, N., and E. Whitchurch. "Mirror, Mirror on the Wall: Enhancement in Self-Recognition." *Personality and Social Psychology Bulletin* 34 (2008): 1159–70.

Guess, A., B. Nyhan, and J. Reifler. "Selective Exposure to Misinformation:

Evidence from the Consumption of Fake News During the 2016 U.S. Presidential Campaign." Unpublished manuscript, Dartmouth College, 2018.

Hall, J. R., E. M. Bernat, and C. J. Patrick. "Externalizing Psychopathology and the Error-Related Negativity." *Psychological Science* 18 (2007): 326–33.

Hardin, C. D., and E. T. Higgins. "Shared Reality: How Social Verification Makes the Subjective Objective." In Richard M. Sorrentino and E. Tory Higgins, eds. *Handbook of Motivation and Cognition, Vol. 3: The Interpersonal Context*. New York: Guilford Press, 1996, pp. 28–84.

Heath, C., C. Bell, and E. Sternberg. "Emotional Selection in Memes: The Case of Urban Legends." *Journal of Personality and Social Psychology* 81 (2001): 1028.

Lieberman, M. D., and N. I. Eisenberger. "The Dorsal Anterior Cingulate Cortex Is Selective for Pain: Results from Large-Scale Reverse Inference." *Proceedings of the National Academy of Sciences* 112 (2015): 15250–55.

Mercier, H., and D. Sperber. "Why Do Humans Reason? Arguments for an Argumentative Theory." *Behavioral and Brain Sciences* 34 (2011): 57–74.

Mischel, W., Y. Shoda, and M. L. Rodriguez. "Delay of Gratification in Children." *Science* 244 (1989): 933.

Moss, F. A., T. Hunt, K. T. Omwake, and M. M. Ronning. *George Washington University Social Intelligence Test*. Washington, DC: Center for Psychological Service, 1925.

Murphy, S. C., F. K. Barlow, and W. von Hippel. "A Longitudinal Test of Three Theories of Overconfidence." *Social Psychological and Personality Science* 9, no. 3 (2017): 353–63.

Murphy, S. C., W. von Hippel, S. L. Dubbs, M. J. Angilletta Jr., R. S. Wilson, R. Trivers, and F. K. Barlow. "The Role of Overconfidence in Romantic Desirability and Competition." *Personality and Social Psychology Bulletin* 41 (2015): 1036–52.

Schlam, T. R., N. L. Wilson, Y. Shoda, W. Mischel, and O. Ayduk. "Preschoolers' Delay of Gratification Predicts Their Body Mass 30 Years Later." *Journal of Pediatrics* 162 (2013): 90–93.

Smith, M. K., R. Trivers, and W. von Hippel. "Self-Deception Facilitates Interpersonal Persuasion." *Journal of Economic Psychology* 63 (2017): 93–101.

Strang, R. "An Analysis of Errors Made in a Test of Social Intelligence." *Journal of Educational Sociology* 5 (1932): 291–99.

Suddendorf, T. *The Gap: The Science of What Separates Us from Other Animals*. New York: Basic Books, 2013.

Tomasello, M., M. Carpenter, J. Call, T. Behne, and H. Moll. "Understanding and Sharing Intentions: The Origins of Cultural Cognition." *Behavioral and Brain Sciences* 28 (2005): 675–91.

von Hippel, W., and K. Gonsalkorale. "'That Is Bloody Revolting!' Inhibitory Control of Thoughts Better Left Unsaid." *Psychological Science* 16 (2005): 497–500.

von Hippel, W., R. Ronay, E. Baker, K. Kjelsaas, and S. C. Murphy. "Quick

Thinkers Are Smooth Talkers: Mental Speed Facilitates Charisma." *Psychological Science* 27 (2016): 119–22.

von Hippel, W., and R. Trivers. "The Evolution and Psychology of Self-Deception." *Behavioral and Brain Sciences* 34 (2011): 1–16.

Wojcik, S. P., A. Hovasapian, J. Graham, M. Motyl, and P. H. Ditto. "Conservatives Report, but Liberals Display, Greater Happiness." *Science* 347 (2015): 1243–46.

6: *HOMO INNOVATIO*

Baron-Cohen, S. "Autism and the Technical Mind." *Scientific American* 307 (2012): 72–75.

Baron-Cohen, S., P. Bolton, S. Wheelwright, L. Short, G. Mead, A. Smith, and V. Scahill. "Does Autism Occur More Often in Families of Physicists, Engineers, and Mathematicians?" *Autism* 2 (1998): 296–301.

Baron-Cohen, S., S. Wheelwright, R. Skinner, J. Martin, and E. Clubley. "The Autism-Spectrum Quotient (AQ): Evidence from Asperger Syndrome/High-Functioning Autism, Males and Females, Scientists and Mathematicians." *Journal of Autism and Developmental Disorders* 31 (2001): 5–17.

Boehm, C. *Hierarchy in the Forest: The Evolution of Egalitarian Behavior.* Cambridge, MA: Harvard University Press, 2009.

Cacioppo, J. T., S. Cacioppo, G. C. Gonzaga, E. L. Ogburn, and T. J. VanderWeele. "Marital Satisfaction and Break-ups Differ Across On-line and Off-line Meeting Venues." *Proceedings of the National Academy of Sciences* 110 (2013): 10135–40.

Gest, S. D., S. A. Graham-Bermann, and W. W. Hartup. "Peer Experience: Common and Unique Features of Number of Friendships, Social Network Centrality, and Sociometric Status." *Social Development* 10 (2001): 23–40.

Giuri, P., M. Mariani, S. Brusoni, G. Crespi, D. Francoz, A. Gambardella . . . and B. Verspagen. "Inventors and Invention Processes in Europe: Results from the PatVal-EU Survey." *Research Policy* 36 (2007): 1107–27.

Gluckman, M., and S. P. Johnson. "Attentional Capture by Social Stimuli in Young Infants." *Frontiers in Psychology* 4 (2013): 527.

Goodall, J. *The Chimpanzees of Gombe: Patterns of Behavior.* Cambridge, MA: Harvard University Press, 1986.

Harari, Y. N. *Sapiens: A Brief History of Humankind.* New York: HarperCollins, 2015.

Hassett, J. M., E. R. Siebert, and K. Wallen. "Sex Differences in Rhesus Monkey Toy Preferences Parallel Those of Children." *Hormones and Behavior* 54 (2008): 359–64.

Lieberman, M. D. *Social: Why Our Brains Are Wired to Connect.* New York: Oxford University Press, 2013.

Lubinski, D., C. P. Benbow, and H. J. Kell. "Life Paths and Accomplishments of Mathematically Precocious Males and Females Four Decades Later." *Psychological Science* 25 (2014): 2217–32.

Lutchmaya, S., and S. Baron-Cohen. "Human Sex Differences in Social and Non-Social Looking Preferences, at 12 Months of Age." *Infant Behavior and Development* 25 (2002): 319–25.

McClure, E. B. "A Meta-analytic Review of Sex Differences in Facial Expression Processing and Their Development in Infants, Children, and Adolescents." *Psychological Bulletin* 126 (2000): 424.

Moss-Racusin, C. A., J. F. Dovidio, V. L. Brescoll, M. J. Graham, and J. Handelsman. "Science Faculty's Subtle Gender Biases Favor Male Students." *Proceedings of the National Academy of Sciences* 109 (2012): 16474–79.

Roelfsema, M. T., R. A. Hoekstra, C. Allison, S. Wheelwright, C. Brayne, F. E. Matthews, and S. Baron-Cohen. "Are Autism Spectrum Conditions More Prevalent in an Information-Technology Region? A School-Based Study of Three Regions in the Netherlands." *Journal of Autism and Developmental Disorders* 42 (2012): 734–39.

Stoet, G., and D. C. Geary. "The Gender-Equality Paradox in Science, Technology, Engineering, and Mathematics Education." *Psychological Science* 29 (2018): 581–93.

Su, R., J. Rounds, and P. I. Armstrong. "Men and Things, Women and People: A Meta-Analysis of Sex Differences in Interests." *Psychological Bulletin* 135 (2009): 859–84.

Suddendorf, T. *The Gap: The Science of What Separates Us from Other Animals*. New York: Basic Books, 2013.

Van Meter, K. C., L. E. Christiansen, L. D. Delwiche, R. Azari, T. E. Carpenter, and I. Hertz-Picciotto. "Geographic Distribution of Autism in California: A Retrospective Birth Cohort Analysis." *Autism Research* 3 (2010): 19–29.

von Hippel, E., J. P. De Jong, and S. Flowers. "Comparing Business and Household Sector Innovation in Consumer Products: Findings from a Representative Study in the United Kingdom." *Management Science* 58 (2012): 1669–81.

Wang, M. T., J. S. Eccles, and S. Kenny. "Not Lack of Ability but More Choice: Individual and Gender Differences in Choice of Careers in Science, Technology, Engineering, and Mathematics." *Psychological Science* 24 (2013): 770–75.

Williams, W. M., and S. J. Ceci. "National Hiring Experiments Reveal 2:1 Faculty Preference for Women on STEM Tenure Track." *Proceedings of the National Academy of Sciences* 112 (2013): 5360–65.

If you are interested in further reading, chapter 6 is based on the following academic paper:

von Hippel, W., and T. Suddendorf. "Did Humans Evolve to Innovate with a Social Rather than Technical Orientation?" *New Ideas in Psychology* 51 (2018): 34–39.

7: ELEPHANTS AND BABOONS

Acemoglu, D., and J. A. Robinson. *Why Nations Fail: The Origins of Power, Prosperity, and Poverty*. New York: Crown Business, 2013.

Archie, E. A., T. A. Morrison, C. A. H. Foley, C. J. Moss, and S. C. Alberts. "Dominance Rank Relationships Among Wild Female African Elephants, *Loxodonta africana*." *Animal Behaviour* 71 (2006): 117–27.

Betzig, L. L. "Despotism and Differential Reproduction: A Cross-Cultural Correlation of Conflict Asymmetry, Hierarchy, and Degree of Polygyny." *Ethology and Sociobiology* 3 (1982): 209–21.

Bidwell, M. "Paying More to Get Less: Specific Skills, Matching, and the Effects of External Hiring Versus Internal Promotion." *Administrative Science Quarterly* 56 (2011): 369–407.

Case, C. R., and J. K. Maner. "Divide and Conquer: When and Why Leaders Undermine the Cohesive Fabric of Their Group." *Journal of Personality and Social Psychology* 107 (2014): 1033–50.

Chagnon, N. A. *Noble Savages: My Life Among Two Dangerous Tribes—the Yanomamö and the Anthropologists*. New York: Simon and Schuster, 2013.

Cowlishaw, G., and R. I. M. Dunbar. "Dominance Rank and Mating Success in Male Primates." *Animal Behavior* 41 (1991): 1045–56.

Inglehart, R., C. Haerpfer, A. Moreno, C. Welzel, K. Kizilova, J. Diez-Medrano, M. Lagos, P. Norris, E. Ponarin, and B. Puranen, et al., eds. "World Values Survey: Wave 6 (2010–2014)." Madrid: JD Systems Institute, 2014. http://www.worldvaluessurvey.org/WVSDocumentationWV6.jsp.

Lawson, D. W., S. James, E. Ngadaya, B. Ngowi, S. G. Mfinanga, and M. B. Mulder. "No Evidence That Polygynous Marriage Is a Harmful Cultural Practice in Northern Tanzania." *Proceedings of the National Academy of Sciences* 112 (2015): 13827–32.

Loughnan, S., P. Kuppens, J. Allik, K. Balazs, S. de Lemus, K. Dumont, R. Gargurevich, I. Hidegkuti, B. Leidner, L. Matos, J. Park., A. Realo, J. Shi, V. Sojo, Y.-Y. Tong, J. Vaes, P. Verduyn, V. Yeung, and N. Haslam. "Economic Inequality Is Linked to Biased Self-Perception." *Psychological Science* 22 (2011): 1254–58.

Maner, J. K., and N. L. Mead. "The Essential Tension Between Leadership and Power: When Leaders Sacrifice Group Goals for the Sake of Self-Interest." *Journal of Personality and Social Psychology* 99 (2010): 482–97.

Marlowe, F. *The Hadza: Hunter-Gatherers of Tanzania*. Berkeley: University of California Press, 2010.

McComb, K., C. Moss, S. M. Durant, L. Baker, and S. Sayialel. "Matriarchs as Repositories of Social Knowledge in African Elephants." *Science* 292 (2001): 491–94.

McComb, K., G. Shannon, S. M. Durant, K. Sayialel, R. Slotow, J. Poole, and C. Moss. "Leadership in Elephants: The Adaptive Value of Age." *Proceedings of the Royal Society B: Biological Sciences* 278 (2001): 3270–76.

Ronay, R., J. K. Oostrom, N. Lehmann-Willenbrock, and M. Van Vugt. "Pride Before the Fall: (Over) Confidence Predicts Escalation of Public Commitment." *Journal of Experimental Social Psychology* 69 (2017): 13–22.

Sapolsky, R. *A Primate's Memoir: A Neuroscientist's Unconventional Life Among the Baboons*. New York: Simon and Schuster, 2007.

Wright, R. *The Moral Animal: Why We Are the Way We Are: The New Science of Evolutionary Psychology*. New York: Pantheon, 1994.

If you are interested in further reading, chapter 7 is based on the following academic paper:

Ronay, R., W. W. Maddux, and W. von Hippel. "Inequality Rules: Resource Distribution and the Evolution of Dominance- and Prestige-Based Leadership." *Leadership Quarterly* (in press).

8: TRIBES AND TRIBULATIONS

Brewer, M. B. "The Psychology of Prejudice: Ingroup Love and Outgroup Hate?" *Journal of Social Issues* 55 (1999): 429–44.

Case, T. I., B. M. Repacholi, and R. J. Stevenson. "My Baby Doesn't Smell as Bad as Yours: The Plasticity of Disgust." *Evolution and Human Behavior* 27 (2006): 357–65.

Chagnon, N. A. "Life Histories, Blood Revenge, and Warfare in a Tribal Population." *Science* 239 (1988): 985.

Cosmides, L., H. C. Barrett, and J. Tooby. "Adaptive Specializations, Social Exchange, and the Evolution of Human Intelligence." *Proceedings of the National Academy of Sciences* 107 (2010): 9007–14.

Fincher, C. L., and R. Thornhill. "Assortative Sociality, Limited Dispersal, Infectious Disease and the Genesis of the Global Pattern of Religion Diversity." *Proceedings of the Royal Society of London B: Biological Sciences* 275 (2008): 2587–94.

———. "A Parasite-Driven Wedge: Infectious Diseases May Explain Language and Other Biodiversity." *Oikos* 117 (2008): 1289–97.

Glowacki, L., and R. Wrangham. "Warfare and Reproductive Success in a Tribal Population." *Proceedings of the National Academy of Sciences* 112 (2015): 348–53.

Keeley, L. H. *War Before Civilization: The Myth of the Peaceful Savage*. New York: Oxford University Press, 1997.

Mathew, S., and R. Boyd. "Punishment Sustains Large-Scale Cooperation in Prestate Warfare." *Proceedings of the National Academy of Sciences* 108 (2011): 11375–80.

Pinker, S. *The Better Angels of Our Nature: The Decline of Violence in History and Its Causes*. London: Penguin UK, 2011.

———. *Enlightenment Now: The Case for Reason, Science, Humanism, and Progress*. New York: Penguin, 2018.

Schaller, M. "The Behavioural Immune System and the Psychology of Human Sociality." *Philosophical Transactions of the Royal Society B: Biological Sciences*, 366 (2011): 3418–26.

Schaller, M., and S. L. Neuberg. "Danger, Disease, and the Nature of Prejudice(s)." *Advances in Experimental Social Psychology* 46 (2012): 1.

Thornhill, R., C. L. Fincher, D. R. Murray, and M. Schaller. "Zoonotic and Non-Zoonotic Diseases in Relation to Human Personality and Societal Values: Support for the Parasite-Stress Model." *Evolutionary Psychology* 8 (2010): 147470491000800201.

Valdesolo, P., and D. DeSteno. "The Duality of Virtue: Deconstructing the Moral Hypocrite." *Journal of Experimental Social Psychology* 44 (2008): 1334–38.

———. "Moral Hypocrisy." *Psychological Science* 18 (2007): 689–90.

Wilson, M. L., and R. W. Wrangham. "Intergroup Relations in Chimpanzees." *Annual Review of Anthropology* 32 (2003): 363–92.

Wilson, R. S., M. J. Angilletta Jr., R. S. James, C. Navas, and F. Seebacher. "Dishonest Signals of Strength in Male Slender Crayfish (*Cherax dispar*) During Agonistic Encounters." *The American Naturalist* 170 (2007): 284–91.

Wood, B. M., D. P. Watts, J. C. Mitani, and K. E. Langergraber. "Favorable Ecological Circumstances Promote Life Expectancy in Chimpanzees Similar to That of Human Hunter-Gatherers." *Journal of Human Evolution* 105 (2017): 41–56.

Wrangham, R. W., and L. Glowacki. "Intergroup Aggression in Chimpanzees and War in Nomadic Hunter-Gatherers." *Human Nature* 23 (2012): 5–29.

Zhang, H., and F. N. von Hippel. "Using Commercial Imaging Satellites to Detect the Operation of Plutonium-Production Reactors and Gaseous-Diffusion Plants." *Science and Global Security* 8 (2000): 261–313.

If you are interested in further reading, chapter 8 is based on the following academic paper:

von Hippel, W. "Evolutionary Psychology and Global Security." *Science and Global Security* 25 (2007): 28–41.

9: WHY EVOLUTION GAVE US HAPPINESS

Baumeister, R. F., E. Bratslavsky, C. Finkenauer, and K. D. Vohs. "Bad Is Stronger than Good." *Review of General Psychology* 5 (2001): 323–70.

Danner, D. D., D. A. Snowdon, and W. V. Friesen. "Positive Emotions in Early Life and Longevity: Findings from the Nun Study." *Journal of Personality and Social Psychology* 80 (2001): 804–13.

Darley, J. M., and C. D. Batson. "'From Jerusalem to Jericho': A Study of Situational and Dispositional Variables in Helping Behavior." *Journal of Personality and Social Psychology* 27 (1973): 100.

Gilbert, D. T., and T. D. Wilson. "Prospection: Experiencing the Future." *Science* 317 (2007): 1351–54.

Kalokerinos, E. K., W. von Hippel, J. D. Henry, and R. Trivers. "The Aging Positivity Effect and Immune Functioning: Positivity in Recall Predicts Higher CD4 Counts and Lower CD4 Activation." *Psychology and Aging* 29 (2014): 636–41.

Kiecolt-Glaser, J. K., T. J. Loving, J. R. Stowell, W. B. Malarkey, S. Lemeshow, S. L. Dickinson, and R. Glaser. "Hostile Marital Interactions, Proinflammatory Cytokine Production, and Wound Healing." *Archives of General Psychiatry* 62 (2005): 1377–84.

Oishi, S., E. Diener, and R. E. Lucas. "The Optimum Level of Well-Being: Can People Be Too Happy?" *Perspectives on Psychological Science* 2 (2007): 346–60.

10: FINDING HAPPINESS IN EVOLUTIONARY IMPERATIVES

Alberts, S. C., J. Altmann, D. K. Brockman, M. Cords, L. M. Fedigan, A. Pusey . . . and A. M. Bronikowski. "Reproductive Aging Patterns in Primates Reveal That Humans Are Distinct." *Proceedings of the National Academy of Sciences* 110 (2013): 13440–45.

Apicella, C. L., F. W. Marlowe, J. H. Fowler, and N. A. Christakis. "Social Networks and Cooperation in Hunter-Gatherers." *Nature* 481 (2012): 497–501.

Bird, R. B., and E. A. Power. "Prosocial Signaling and Cooperation Among Martu Hunters." *Evolution and Human Behavior* 36 (2015): 389–97.

Buss, D. M. "The Evolution of Happiness." *American Psychologist* 55 (2000): 15–23.

———. "Sex Differences in Human Mate Preferences: Evolutionary Hypotheses Tested in 37 Cultures." *Behavioral and Brain Sciences* 12 (1989): 1–14.

Corder, E. H., A. M. Saunders, W. J. Strittmatter, D. E. Schmechel, P. C. Gaskell, G. W. Small, A. D. Roses, J. L. Haines, and M. A. Pericak-Vance. "Gene Dose of Apolipoprotein E Type 4 Allele and the Risk of Alzheimer's Disease in Late Onset Families." *Science* 261 (1993): 921–23.

Easterlin, R. "Will Raising the Incomes of All Increase the Happiness of All?" *Journal of Economic Behaviour and Organization* 27 (1994): 35–47.

Hoffman, M., E. Yoeli, and M. A. Nowak. "Cooperate Without Looking: Why We Care What People Think and Not Just What They Do." *Proceedings of the National Academy of Sciences* 112 (2015): 1727–32.

Hunt, J., R. Brooks, M. D. Jennions, M. J. Smith, C. L. Bentsen, and L. F. Bussiere. "High-Quality Male Field Crickets Invest Heavily in Sexual Display but Die Young." *Nature* 432 (2004): 1024–27.

Inglehart, R., C. Haerpfer, A. Moreno, C. Welzel, K. Kizilova, J. Diez-Medrano, M. Lagos, P. Norris, E. Ponarin, and B. Puranen, et al., eds. "World Values Survey: Wave 6 (2010–2014)." Madrid: JD Systems Institute, 2014. http://www.worldvaluessurvey.org/WVSDocumentationWV6.jsp.

Morgan, D., K. A. Grant, H. D. Gage, R. H. Mach, J. R. Kaplan, O. Prioleau, S. H. Nader, N. Buchheimer, R. L. Ehrenkaufer, and M. A. Nader. "Social Dominance in Monkeys: Dopamine D2 Receptors and Cocaine Self-Administration." *Nature Neuroscience* 5 (2002): 169–74.

Oishi, S., and U. Schimmack. "Residential Mobility, Well-Being, and Mortality." *Journal of Personality and Social Psychology* 98 (2010), 980–94.

Owens, I. P. "Sex Differences in Mortality Rate." *Science* 297 (2002): 2008–9.

Pellegrini, A. D., and P. K. Smith, eds. *The Nature of Play: Great Apes and Humans*. New York: Guilford Press, 2005.

Rand, D. G., J. D. Greene, and M. A. Nowak. "Spontaneous Giving and Calculated Greed." *Nature* 489 (2012): 427–30.

Ronay, R., and W. von Hippel. "The Presence of an Attractive Woman Elevates Testosterone and Physical Risk-Taking in Young Men." *Social Psychological and Personality Science* 1 (2010): 57–64.

van Boven, L., and T. Gilovich. "To Do or to Have? That Is the Question." *Journal of Personality and Social Psychology* 85 (2003): 1193–302.

Wilder, J. A., Z. Mobasher, and M. F. Hammer. "Genetic Evidence for Unequal Effective Population Sizes of Human Females and Males." *Molecular Biology and Evolution* 21 (2004): 2047–57.

If you are interested in further reading, chapters 9 and 10 are based on the following academic paper:

von Hippel, W., and K. Gonsalkorale. "Evolutionary Imperatives and the Good Life." In J. P. Forgas and R. Baumeister, eds. *The Social Psychology of Living Well.* New York: Psychology Press, 2018.

EPILOGUE

Dawkins, R. *Unweaving the Rainbow: Science, Delusion, and the Appetite for Wonder.* New York: Houghton Mifflin Harcourt, 2000.

Sapolsky, R. *A Primate's Memoir: A Neuroscientist's Unconventional Life Among the Baboons*. New York: Simon and Schuster, 2007.

Tomasello, M. *A Natural History of Human Morality*. Cambridge, MA: Harvard University Press, 2016.

Index

Page numbers of illustrations appear in *italics.*

About the Author

WILLIAM VON HIPPEL grew up in Alaska, got his BA at Yale and his PhD at the University of Michigan, and then taught for a dozen years at Ohio State University before finding his way to Australia, where he is a professor of psychology at the University of Queensland. He has published more than a hundred articles and chapters, and his research has been featured in the *New York Times*, *USA Today*, the *Economist*, the BBC, *Le Monde*, *El Mundo*, *Der Spiegel*, and the *Australian*. He lives with his wife and two children in Brisbane, Australia.